MW01032338

Waterproofing the Building Envelope

Waterproofing the Building Envelope

Michael T. Kubal

McGraw-Hill, Inc.
New York St. Louis San Francisco Auckland Bogotá
Caracas Lisbon London Madrid Mexico City Milan
Montreal New Delhi San Juan Singapore
Sydney Tokyo Toronto

Library of Congress Cataloging-in-Publication Data

Kubal, Michael T.
Waterproofing the building envelope / Michael T. Kubal.
p. cm.
Includes index.
ISBN 0-07-035859-1
1. Waterproofing. I. Title.
TA9031.K79 1993
693'.892—dc20 92-12795
CIP

3 4 5 6 7 8 9 10 11 12 13 14 BKMBKM 9 9 8 7 6 5

ISBN 0-07-035859-1

The sponsoring editor for this book was Larry Hager, the editing supervisor was Ruth W. Mannino, and the production supervisor was Suzanne W. Babeuf. This book was set in Century Schoolbook by Carol Woolverton, Lexington, Mass., in cooperation with Warren Publishing Services.

To the women in my life

my mother ***Rose***
my wife **Carole**
my daughter ***Jennifer***

Contents

Preface

Waterproofing the Building Envelope, the first book of its kind, is a pursuit for standards and excellence within the waterproofing and construction industry. Only when waterproofing is acknowledged as a necessary performance standard for all building envelope components will the construction industry begin to resolve problems at job sites instead of in courtrooms.

Waterproofing, often rated first among the most frequent causes of construction complaints, continually creates unnecessary problems. No single building component producing such enormous problems has so little standardization on which to base improvements. The construction industry has yet to adopt the principle that the entire exterior facade must be treated as a single cohesive unit—an envelope in which all individual components are transitioned into one another in completely waterproof detailing.

Such a task is complicated by job-site manufacturing. Even if all facade components are factory-manufactured at the job site, they must be assembled into a single unit. It is here that the industry fails. Waterproofing standards must be developed to transition the various building facade components coherently.

Architects or engineers provide insufficient detailing to transition components together. Contractors lack sufficient supervision to ensure that terminations and transition details are installed properly by multiple trade contractors. Owners are remiss in implementing effective exterior maintenance programs, focusing on building exterior maintenance only when leaks appear.

The installation process must catch up with technology advances in materials and systems. These manufactured products do not produce leakage; it is rather the improper detailing or lack thereof in integrating products into a structure that causes the problem. Waterproofing manufacturers must also become more responsive in providing information on the correct details for transitions and termination methods used to incorporate their materials into a composite building envelope.

Finally, testing must be promoted to ensure the most effective life cycling of these transition components, which are often the initial source of water infiltration and structural damage. What good is a 10-year lifetime for waterproofing membrane when the transition detailing between it and adjoining systems will last only 5 years?

This book is written through actual experience since too little documented information, particularly uniform standards, is available in the waterproofing industry. Evidence the multiple trade unions claiming jurisdiction over individual waterproofing components (e.g., masons for sealants, roofers for membrane installations). Evidence testing laboratories incorporating waterproofing tests into other building components such as roofing. Evidence the inexperienced trades attempting installation of envelope components including drywall contractors installing exterior sheeting and through-wall flashing.

All exterior facade contractors, from the mason installing flashing to the electrician installing exterior lighting, require waterproofing skills. All trades must be made aware of envelope waterproofing requirements and their part in maintaining the watertight effectiveness of a building skin.

This instruction must become part of any total quality management (TQM) program for the construction industry. Only then will the process of cohesively putting together all components of the envelope begin to mature.

This approach must also be taught in the classroom. Waterproofing and building envelope detailing are rarely given appropriate coverage in the learning place. This lack continues to promote the belief that these issues are insignificant in the construction process.

This issue is firmly supported by the legal profession for the construction industry.

This book covers all available waterproofing materials and systems, as well as emphasizes the relationships these systems have to other envelope components. Waterproofing must be considered an integral part of all envelope systems. This includes roofing, masonry, electrical mechanical, and all other systems comprising the envelope. Only then will the construction industry begin to see a reduction in lawsuits and the end of the adversarial relationship that occurs among owners, contractors, architects, and trade contractors.

Michael T. Kubal

Waterproofing the Building Envelope

Chapter

1

The Building Envelope

Since our beginnings, we have sought shelter as protection from the elements. Yet, even today, after centuries of technological advances, we are still confronted by nature's elements contaminating our constructed shelters. This is not due to a lack of effective waterproofing products but rather to the increasing complexity of shelter construction and an inability to coordinate interfacing between the multitude of construction systems involved in a single building.

Adequately controlling groundwater, rainwater, and surface water will prevent damage and avoid unnecessary repairs to building envelopes. In fact, water is the most destructive weathering element of concrete, masonry, and natural stone structures. Waterproofing techniques preserve a structure's integrity and usefulness through an understanding of natural forces and their effect during life cycling. Waterproofing also involves choosing proper designs and materials to counter the detrimental effects of these natural forces.

Site construction requires combining numerous building trades and systems into a building skin to prevent water infiltration. Our inability to tie together these various components effectively causes the majority of water and weather intrusion problems. Experience has shown that as much as 90 percent of water intrusion problems occurs within 1 percent of a building's exterior surface area. An inability to control installation and details, linking various waterproofing systems on a building's exterior skin, creates these problems.

To prevent all possible water intrusion causes, a building must be *enveloped* from top to bottom with waterproof materials. These waterproof systems must then interact integrally to prevent water infiltration. Should any one of these systems fail or not act integrally with all other envelope systems, leakage will occur.

Even with continual technological advances in materials, water continues to create unnecessary problems. This is most often due to an envelope's inability to act as an integrated system preventing water and pollutant infiltration. All too often several systems are designed into a building, chosen independently and acting independently rather than cohesively.

Detailing transitions from one system to another or terminations into structural components are often overlooked. Product substitutions that do not act integrally with other specified systems create problems and leakage. Inadequate attention to movement characteristics of a structure can cause stress to in-place systems that they are not able to withstand. All these situations acting separately, or in combination, will eventually cause water intrusion.

To understand the complete enveloping of a structure, several definitions as well as their relationship to one another must be made clear:

Roofing. That portion of a building that prevents water intrusion in horizontal or slightly inclined elevations. Although typically applied to the surface and exposed to the elements, roofing systems can also be internal, or sandwiched, between other building components.

Below-grade waterproofing. Materials that prevent water under hydrostatic pressure from entering into a structure or its components. These systems are not exposed or subjected to weathering such as by ultraviolet rays.

Above-grade waterproofing. A combination of materials or systems that prevent water intrusion into exposed structure elements. These materials are not subject to hydrostatic pressure but are exposed to weathering and pollutant attack.

Dampproofing. Materials resistant to water vapor or minor amounts of moisture that act as back-up systems to primary waterproofing materials. Dampproofing materials are not subject to weathering or water pressure.

Flashing. Materials or systems installed to redirect water entering through the building skin back to the exterior. Flashings are installed as backup systems for waterproofing or dampproofing systems.

Building envelope. The combination of roofing, waterproofing, dampproofing, and flashing systems that act cohesively as a barrier, protecting interior areas from water and weather intrusion. These systems envelop a building from top to bottom, from below grade to the roof.

The entire exterior building skin must be enveloped to prevent water infiltration. Buildings must be made waterproof from roof coverings to vertical above-grade walls to below-grade floors.

An envelope may contain combinations of several systems, such as a building's facing material (e.g., brick facade), a water repellent, a cavity wall dampproofing, and a through-wall flashing. Each system acts integrally with the others as a total system for complete effectiveness as a weather-tight building envelope. Figure 1.1 shows the interrelationships of various waterproof systems.

In Fig. 1.1, horizontal membrane roofing terminates into a vertical parapet at the metal counterflashing. Parapet wall waterproofing also ties into roofing at this counterflashing. This flashing detail ensures envelope water tightness. It ties two separate systems together so that they may act cohesively.

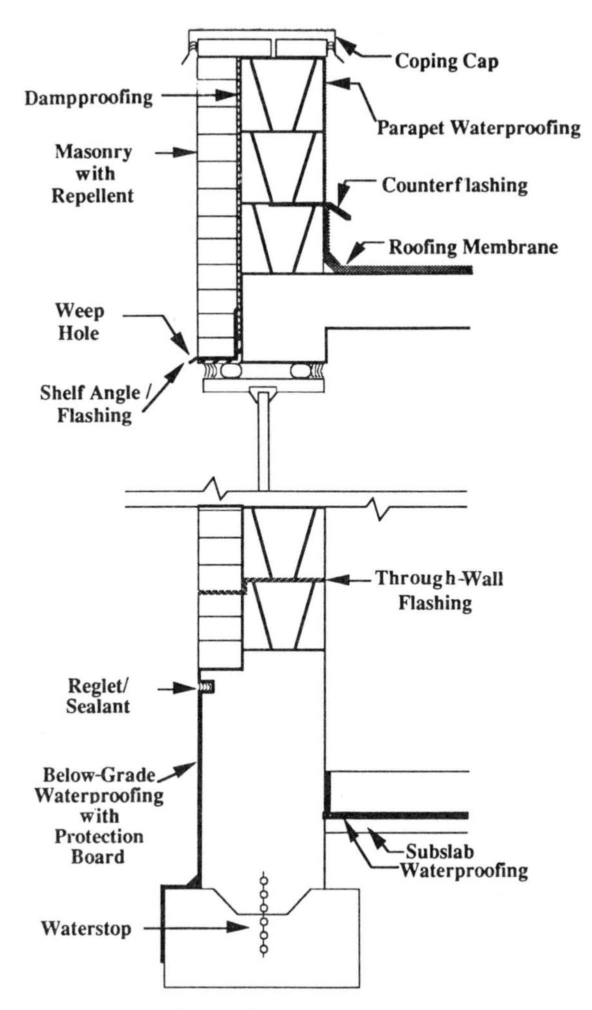

Figure 1.1 Detailing of the building envelope.

A similar detail occurs at the coping cap. This flashing details terminations of parapet waterproofing, building facing material, clear water repellent on brick, and cavity wall dampproofing. Without such critical detail, the independent systems mentioned above could not function cohesively to provide building envelope watertightness.

On the vertical facade, detailing control joints with sealant at the brick facade allows for thermal movement, while maintaining a watertight joint. The brick has through-wall flashing diverting intruding water, collected by dampproofing, back out through weep holes. Additionally, sealant at window perimeters acts as a transition between brick facing and window frames.

Dampproofing terminates into below-grade waterproofing at a reglet placed into the concrete wall. Sealant acts as a transition between these systems. Waterstops tie vertical waterproofing to horizontal slab waterproofing, completing a watertight seal. Drainage systems then channel collected water at foundations to prevent unnecessary water pressure against a building foundation wall.

As shown, each separate waterproofing material effectively joins together to form the building envelope.

Each trade contractor's work, regardless of its being thought of as waterproofing, must become part of a totally watertight building envelope. Equally important, all individual envelope systems must be adequately transitioned into other components or provided with watertight terminations. Often the tradesworkers completing this work are not aware, trained, or supervised in envelope waterproofing. The resulting improper attention to details is responsible for countless problems in construction. Properly detailing a building's envelope presents an enormous task. From inception to installation, numerous stumbling blocks occur. Highlighting this interrelationship of various envelope systems is the following principle:

As much as 90 percent of all water intrusion problems occur within 1 percent of the total building exterior surface area.

The 1 percent of a structure's facade contains the terminations and transition detailing that all too frequently lead to envelope failures. Not until our ability to train and supervise field construction reaches the technology of waterproofing systems and construction techniques will this situation correct itself. It is not the actual manufactured waterproofing systems or envelope systems that leak but the field construction details of terminations and transitions.

Besides preventing water infiltration, waterproofing systems prevent structural damage to building components. In northern climates, watertightness prevents spalling of concrete, masonry, or stone due to freeze–thaw cycles. Watertightness also prevents rusting and deterioration of structural or reinforcing steel encased in exterior concrete or behind masonry facades.

Waterproofing also prevents the passage of pollutants that cause steel deterioration and concrete spalling, such as chloride ion (salts), into structural components. This is especially true in horizontal exposed areas such as balcony decks and parking garages. Prevention of acid rain contamination (sulfites mixing with water to form sulfuric acid) and carbon acids (vehicle exhaust–carbon dioxide that forms carbonic acid when mixed with water) is also an important consideration when choosing proper waterproofing applications.

Building envelopes also provide energy savings and environmental control by acting as weather barriers against wind, cold, and heat. Additionally, envelopes must be resistant to wind loading and wind infiltration. These forces, in combination with water, can multiply the magnitude of damage to a structure and its interior contents. Direct wind load pressure can force water deeper into a structure through cracks or crevices, where water might not normally penetrate. It also creates vertical upward movement of water over window sills and through vents and louvers. Air pressure differences due to wind conditions may cause water that is present to be sucked into a structure because of negative pressure in interior areas.

This situation occurs when outside air pressure is greater than interior air pressure. It also occurs through a churning effect, where cool air is pulled into lower portions of a building, replacing warmer air that rises and escapes through higher areas. To prevent this forced water infiltration and associated energy loss, a building envelope must be resistant and weather tight against wind.

For envelopes to function as intended requires proper attention to

- Selection and design of compatible materials and systems
- Proper detailing of material junctions and terminations
- Installation and inspection of these details during construction
- Ability of composite envelope systems to function during weathering cycles
- Maintenance by building owners

From the numerous systems available to a designer, specific products that can function together must be chosen. Once products are chosen and specified, proposed substitutions by contractors must be carefully reviewed. Similar products may not function or be compatible with previously chosen components. Substitutions with multiple systems of specified components only further complicate matters.

Improper attention to specified details of terminations, junctures, and changes in materials during installation can cause water infiltration. Once construction begins, installation procedures must be continually monitored to meet specified design criteria and manufacturers' recommendations. Detailing problems compound by using several different crafts and subcontractors in a single detail. For instance, a typical coping cap detail involves roofing, carpentry, masonry, waterproofing, and sheet metal contractors. One weak or improperly installed material in this detail will create problems for the entire envelope.

Finally, products chosen and installed as part of a building envelope must function together during life cycling and weathering of a structure. Proper maintenance after system installation is imperative for proper life cycling. Will shelf angles be adequate to support parapet walls during wind or snow weathering? Will oxidation of counterflashing allow water infiltration into a roof system causing further deterioration?

From these processes of design, construction, and maintenance, 99 percent of a building envelope will typically function properly. It is the remaining 1 percent that creates the magnitude of problems. This 1 percent requires much more attention and time by owners, architects, engineers, contractors, and subcontractors to ensure an effective building envelope.

The most frequent problems of this 1 percent occur because of inadequate design detailing by architects, improper installation by contractors, and improper maintenance by owners. Typical frequent errors include

Architect and engineer. Improper detail specifications; no allowance for structural or thermal movement; improper selection of materials; use of substitutes that do not integrate with other components of the envelope

Contractor and subcontractor. Improper installations; inattention to details; no coordination of the subcontractors; use of untrained mechanics to complete the work

Building owners and managers. No scheduled maintenance programs; use of untrained personnel make repairs; no scheduled inspection programs; postponement of repairs until further damage is done

The chapters in this book review each component of a waterproofing envelope in detail. They are followed by a chapter on bringing individual systems together for a watertight building envelope. In addition to the information presented, standard test results completed in accordance with the National Bureau of Standards (NBS) (Federal Standards) and the American Society for Testing and Materials (ASTM) are referred to. These results from standardized tests allow comparison of various materials and systems to ensure that proper envelope materials are chosen for a particular installation. Refer to Chap. 10 for information on testing.

Testing also allows material selection that ensures compatibility with other integral envelope systems. Remember, no matter how good a material is chosen, a building envelope will succeed only if all components act as a cohesive unit. One weak detail on an envelope can cause the entire system to fail.

Chapter

2

Below-Grade Waterproofing

Water in the form of vapor, liquid, and solids presents below-grade construction with many unique problems. Water causes damage by vapor transmission through porous surfaces, by direct leakage in a liquid state, and by spalling of concrete floors in a frozen or solid form. Water conditions below grade make interior spaces uninhabitable not only by leakage but also by damage to structural components as exhibited by reinforcing steel corrosion, concrete spalling, settlement cracks, and structural cracking.

Surface Water Control

Water present at below-grade surfaces is available from two sources—surface water and groundwater. Beyond selection and installation of proper waterproofing materials, all waterproof installations must include methods for control and drainage of both surface and groundwater.

Surface water from sources including rain, sprinklers, and melting snow should be directed immediately away from a structure. This prevents percolation of water directly adjacent to perimeter walls or water migration into a structure. Directing water is completed by one or a combination of steps. Soil adjacent to a building should be graded and sloped away from the structure. Slopes should be a minimum of ½ in/ft for natural areas, paved areas, and sidewalks sloped positively to drain water away from the building.

Automatic sprinklers directed against building walls can saturate above-grade walls causing leakage into below-grade areas. Downspouts or roof drains, as well as trench drains installed to direct large amounts of water into drains, direct water away from a building.

Groundwater Control

Besides protection from normal groundwater levels, allowance is made for temporary rises in groundwater levels to protect interior areas. Groundwater levels rise due to rain accumulations and natural capillary action of soils. Waterproof materials must be applied in heights sufficient to prevent infiltration during temporarily raised groundwater levels.

With every below-grade installation, a system for collecting, draining, and discharging water away from foundations is recommended. Foundation drains are effective means for proper collection and discharge. They consist of a perforated pipe installed with perforations set downward in a bed of gravel that allows water drainage. Perforated drain piping is usually polyvinyl chloride (PVC). Vitreous clay piping is sometimes used, but it is more susceptible to breakage. Drain piping is installed next to and slightly above the foundation bottom, not below foundation level to prevent the washing away of soil under the foundation that can cause structural settlement.

Coarse gravel is installed around and over drainage piping for percolation and collection of water. Additionally, mesh or mats are installed over the gravel to prevent soil buildup, which can seal drainage piping perforations and prevent water drainage. Collected water must be drained by natural sloping of pipe to drain fields or pumped into sump pits.

Drainage mats are available for vertical drainage against wall surfaces. These are applied directly against waterproof membranes. These mats have a thick open weave providing downward drainage into drainage fields and prevent water from coming in direct contact with waterproof systems. Drainage mats vary in thickness, but are approximately 1 in thick. Some material manufacturers prefer their use over applied waterproofing systems in place of protection board.

Additionally, waterstops should be used at all "cold joints" on foundation walls. These joints occur where separate pours of concrete meet after one pour has cured. Waterstop installations include wall-to-foundation joints, wall-to-wall intersections, and floor-to-wall joints. Waterstops are manufactured from a variety of materials including PVC.

Waterstops should be continuous in a joint, with no laps or seams used. After placement of waterstop in the first half of the pour, the waterstop must be properly positioned into the second half and not allowed to be folded over. This would negate its function and create voids in the concrete that could crack and allow direct water infiltration.

Construction details must be included to prevent natural capillary action of soils beneath foundations or below-grade floors. Capillary action is upward movement of water and vapor through voids in soil from wet lower areas to drier high areas. Capillary action is dependent upon the soil type present. Clay soils promote the most capillary action, allowing more than 10 ft of vertical capillary action. Loose coarse gravel prevents capillary action, with this type of soil promoting virtually no upward movement.

Capillary action begins by liquid water saturating lower areas adjacent to the water source. This transgresses to a mixture of liquid and vapor above the saturation layer. Finally, only vapor exists in upper soil areas. This vapor is as damaging as water to interior building areas.

Water vapor penetrates pores of concrete floors, condensing into water once it reaches adjacent air-conditioned space. This condensation causes delamination of finished floor surfaces, mildew, and staining. Therefore, it is necessary to prevent or limit capillary action, even when using waterproof membranes beneath slabs. Excavating sufficiently below finished floor elevation and installing a bed of capillary-resistant soil provides drainage of water beneath slabs on grade.

This combination of foundation drainage and soil composition directs water away from a structure and is necessary for any waterproofing and envelope installation. Recommended controls for proper surface and groundwater control are summarized in Fig. 2.1.

Below-grade waterproofing prevents water under hydrostatic pressure from entering into a structure or its components. Below-grade products are not exposed or subjected to weathering by ultraviolet rays or thermal movement. Therefore, unlike above-grade products, they are not inherently resistant to these elements. Below-grade waterproofing systems require application of a protection layer course during backfilling or concrete placement operations.

Positive and Negative Systems

In new and remedial installations, there are both *negative side* and *positive side* below-grade systems. Positive side waterproofing applies to sides with direct exposure to water or a hydrostatic head of water. Negative side waterproofing applies to the opposite or interior side from which water occurs. Examples are shown in Fig. 2.2.

Although both systems have distinct characteristics, as summarized in Table 2.1, the majority of available products are positive-type systems. Negative systems are limited to cementitious based materials, which are frequently used for remedial applications. Some materials apply to negative sides of a structure for remedial applications but function as positive side waterproofing. These materials include chemical grouts, epoxy grouts, and pressure grouts. Admixtures (material

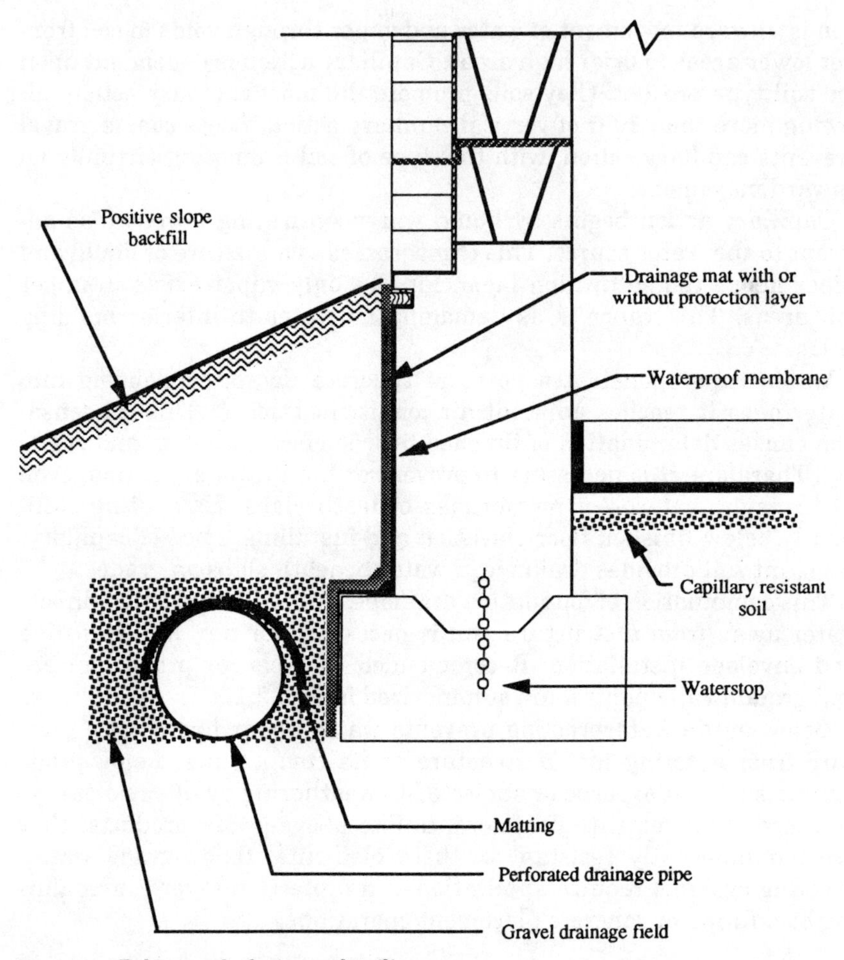

Figure 2.1 Below-grade drainage detailing.

added or mixed into mortars, plaster, stucco, and concrete) have both positive and negative features but are not as effective as surface applied systems.

The principal advantage of a negative system is also its principal disadvantage. It allows water to enter a concrete substrate, promoting both active curing and the corrosion and deterioration of reinforcing steel if chlorides are present. Positive side waterproofing produces an opposite result—no curing of concrete, but protection of reinforcing steel, and of the substrate itself.

Positive and negative below-grade systems include

- Cementitious systems
- Fluid-applied membranes

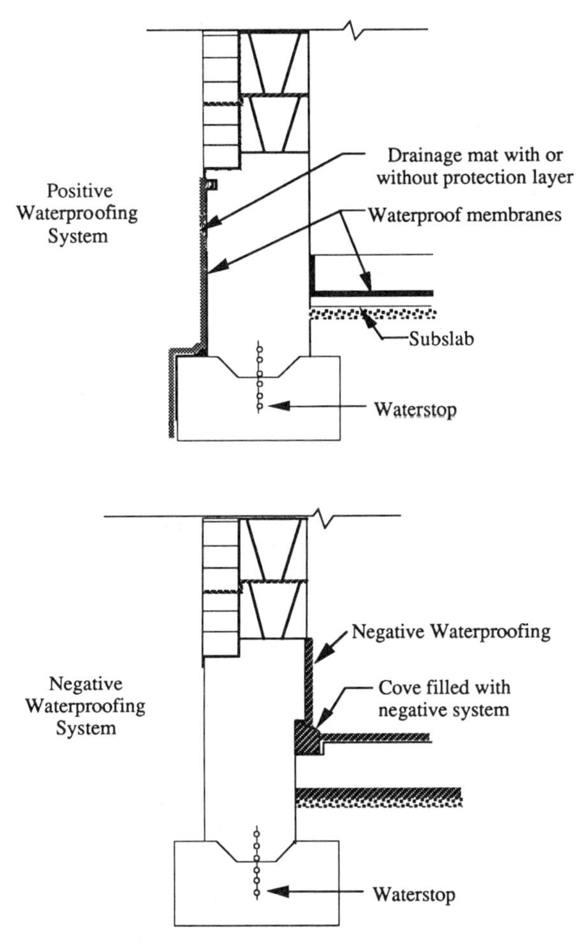

Figure 2.2 Below-grade positive and negative waterproofing details.

- Sheet-membrane systems
- Hydros clay
- Vapor barriers

Cementitious Systems

Cementitious waterproofing systems contain a base of Portland cement, with or without sand, and an active waterproofing agent. There are four types of cementitious systems: metallic, capillary system, chemical additive systems, and acrylic modified systems. Cementitious systems are effective in both positive and negative applications, as well

TABLE 2.1 Comparison of Positive and Negative Waterproofing Systems*

Positive systems	Negative systems
Advantages	
Water is prevented from entering substrate surface	Accessible after installation
Substrate protected from freeze–thaw cycles	Concrete substrate is allowed to moist cure
Substrate is protected from corrosive chemicals in groundwater	Eliminates need for subslabs and well pointing for foundation waterproofing
Disadvantages	
Concrete may not cure properly	Limited to application of cementitious systems
System inaccessible for repairs after installation	No protection from freeze–thaw cycles
Subslabs and well pointing necessary for foundation waterproofing	No protection of substrate or reinforcing steel from groundwater and chemicals

*With all positive systems, concrete should cure properly (up to 21 days) before application of any waterproofing materials.

as in remedial applications. These systems are brushed or troweled to concrete or masonry surfaces and become an integral part of a substrate.

In new construction, where costs and scheduling are critical, these systems are particularly effective. They do not require a completely dry substrate, and concrete does not need to be fully cured before application. This eliminates well pointing and the need for water control during construction. These systems apply to both walls and floors at one time, thereby eliminating staging of waterproofing operations. No subslabs are required for horizontal applications in new construction preventative waterproofing installations. Finally, in cases such as elevator pits, the waterproofing is completed almost any time during construction as best fits scheduling.

All cementitious systems are similar in application and performance but repel water differently by the proprietary additives of a manufacturer's formulations. Cementitious systems have several mutual advantages, including seamless application after which no protection board installation is necessary.

All cementitious materials lack crack bridging or elastomeric properties but are successfully applied to below-grade areas that do not experience thermal movement. However, below-grade areas are subject to freeze–thaw cycling and structural settlement. If these cause movement or cracking, a cementitious system will crack, allowing water infiltration.

Metallic systems

Metallic materials contain a mixture of sand and cement with finely graded iron aggregate or filings. When mixed with water to form a slurry for application, the water acts as an agent permitting the iron filings to oxidize. These materials expand due to this oxidizing, which then effectively seals a substrate and prohibits further transmission of water through the material. This system is one of the oldest methods used for waterproofing and remains an effective waterproofing system. (See Fig. 2.3.)

Metallic systems are applied in two or three coats with the final coat a sand and cement mixture providing protection over base coat waterproofing where exposed. This final coat seals the metallic coats and prevents leaching or oxidization through paints or finishes applied over waterproofed areas. To prevent excessive wear, concrete toppings are installed over horizontal exposed surfaces subject to pedestrian or vehicular traffic.

If drywall or paneling is installed over the waterproofing, furring strips are first applied by gluing them directly to the cementitious system. This eliminates nailing the boards through the cementitious membrane. Carpet perimeter tracks should be applied in the same manner to prevent damage.

Figure 2.3 Negative application of cementitious (metallic) waterproofing. (*Courtesy of Western Group*)

Capillary systems

Capillary cementitious materials are formulations using combinations of proprietary chemicals in dry or liquid form with sand and cement. When applied to a substrate, active chemicals in the slurry react with free lime and moisture in the substrate. This reaction produces crystalline growth in concrete or masonry substrate capillaries. These crystals fill open pores, effectively blocking the passage of water. As with metallic systems, these capillary systems become an integral part of a substrate.

Curing installed systems is critical for adequate crystalline growth. The curing should continue 24–48 hours after installation. Concrete or masonry substrates must be wet to apply these systems, which may be installed over uncured concrete.

In exposed interior applications, coating installation should be protected by plastic, drywall, or paneling applied over furring strips. Floor surfaces are protected by concrete overlays, carpet, or tile finishes.

Chemical additive systems

Chemical cementitious systems are a mixture of sand, cement, and proprietary chemicals (inorganic or organic), which when applied to masonry or concrete substrates provide a watertight substrate by chemical action. Proprietary chemicals are unique to each manufacturer, but typically include silicate and siloxane derivatives in combination with other chemicals. These systems also become an integral part of a substrate after application.

Chemical cementitious systems, approximately 1/16 in thick, are thinner applications than other cementitious products. As with all cementitious systems, concrete or masonry substrates need not be dry for application. Chemical systems do not require curing, but capillary systems do.

Acrylic modified systems

Acrylic modified cementitious systems add acrylic emulsions to a basic cement and sand mixture. These acrylics add waterproofing characteristics and properties to in-place materials. Acrylic systems are applied in two trowel applications, with a reinforcing mesh added into the first layer immediately upon application. This mesh adds some crack bridging capabilities to acrylic installations. However, since the systems bond tenaciously to concrete or masonry substrates, movement capability is limited.

Acrylic cementitious systems are applicable with both positive and negative installations. Concrete substrates can be damp but must be cured for acrylic materials to bond properly. Alkaline substrates can deter performance of acrylic modified cementitious systems.

Acrylic modified materials are applied in a total thickness of approximately ⅛ in. Reinforcing mesh eliminates the need for protective covering of the systems on floor areas in minimal or light traffic interior areas.

The properties of all types of cementitious systems are summarized in Table 2.2.

Cementitious System Application

Before applying cementitious systems substrates must be free of dirt, laitance, form release agents, and other foreign materials. Manufacturers typically require concrete surfaces to be acid etched, lightly sand blasted, or bush hammered to a depth of cut of approximately 1⁄16 in. This ensures adequate bonding to a substrate.

All tie holes, honeycomb, and cracks must be filled by packing them with an initial application of the cementitious system. Construction joints, wall-to-floor joints, wall-to-wall intersections, and other changes in plane should be formed or grooved with a 1-in by 1-in cutout to form a cove. This cove is then packed with cementitious material before initial application. This is a critical detail for cementitious systems as they do not allow for structural or thermal movement. This cove prevents water infiltration at weak points in a structure where cracks typically develop. At minimum, if a cove is not formed, place a cant of material at the intersections, using a dry mix of cementitious material.

Cementitious systems do not require priming of a substrate before application. However, wetting of the concrete with water is necessary. (See Fig. 2.4.)

Cementitious systems are available in a wide range of packaging. They may be premixed with sand and cement in pails, or chemicals and iron may be provided in separate containers and added to the sand and cement mixture. Products are mixed in accordance with manufacturers' recommendations, adding only clean water.

Typically, cementitious systems are applied in two coats after the ini-

TABLE 2.2 Properties of Cementitious Waterproofing Systems

Advantages	Disadvantages
Positive or negative applications	No movement capability
Remedial applications	Job-site mixing required
No subslabs or well pointing required	Not for high traffic areas

Figure 2.4 Surface preparation for waterproofing application. (*Courtesy of Western Group*)

tial preparatory work is complete. First coats may be proprietary materials only. Second coats are usually the chemical or metallic materials within a cement and sand mixture. Third coats are applied if additional protection is necessary. They consist only of sand and cement for protecting exposed portions or adding texture. Acrylic systems often require a reinforcing mesh to be embedded into the first coat application.

Thickness of a system depends upon the sand and cement content of the coatings. The systems are applied by trowel, brush, or spray. Certain systems are dry broadcast over just-placed concrete floors to form a waterproofing surface integral with the concrete.

This method is referred to as the dry-shake application method. Broadcasting powder onto green concrete is followed by power troweling to finish the concrete and distribute the chemicals that are activated by the concrete slurry. This method should not be used for critical areas of a structure subject to water head, as it is difficult to monitor and control.

To protect exposed floor applications, a 2-in concrete topping, carpet, tile, or other finish is applied over the membrane. Walls can be finished with a plaster coating or furred out with adhesively applied drywall or other finish systems.

These systems require proper curing of the cementitious waterproof coating, usually a wet cure of 24–48 hours. Some systems may have a chemical additive to promote proper curing.

These systems do not withstand thermal or structural substrate movement. Therefore, they require special detailing at areas that are experiencing movement such as wall–floor intersections. It is advantageous to install negative cementitious systems after a structure is completely built. This allows structural movement such as settling to occur before application.

Fluid-Applied Systems

Fluid-applied waterproof materials are solvent-based mixtures containing a base of urethanes, rubbers, plastics, vinyls, polymeric asphalts, or combinations thereof. Fluid membranes are applied as a liquid and cure to form a seamless sheet. Since they are fluid applied, controlling thickness is critical during field application (see Fig. 2.5).

Therefore, field measurements must be made (wet or dry film) for millage control. The percentages of solids in uncured material vary. Those with 75 percent solids or less can shrink, causing splits, pinholes, or insufficient millage to waterproof adequately.

Fluid systems are positive waterproofing side applications and require a protection layer before backfilling. Fluid-applied systems are frequently used because of their ease of application, seamless curing, and adaptability to difficult detailing, such as penetrations and changes in plane. These systems allow both above- and below-grade applications, including planters and split-slab construction. Fluid systems are not resistant to ultraviolet weathering and cannot withstand foot traffic and, therefore, are not applied at exposed areas.

There are several important installation procedures that must be followed to ensure performance of these materials. These include proper concrete curing (minimum 7 days, 21–28 days preferred), dry and clean substrate, and proper millage. Should concrete substrates be wet, damp, or uncured, fluid membranes will not adhere and blisters will

Figure 2.5 Spray application of fluid-applied membrane. (*Courtesy of Western Group*)

occur. Proper thickness and uniform application are important for a system to function as a waterproofing material.

Materials can be applied to both vertical and horizontal surfaces, but with horizontal applications a subslab must be in place so that the membrane can be applied to it. A topping, including tile, concrete slabs, or other hard finishes, is then applied over the membrane. Fluid materials are applicable over concrete, masonry, metal, and wood substrates.

Fluid-applied systems have elastomeric properties with tested elongation over 500 percent, with recognized testing such as ASTM C-836. This enables fluid-applied systems to bridge substrate cracking up to 1⁄16 in wide.

An advantage with fluid systems is their self-flashing installation capability. This application enables material to be applied seamless at substrate protrusions, changes in planes, and floor–wall junctions. Fluid materials are self-flashing with no other accessories required for transitions into other building envelope components. However, a uniform 50–60 mil is difficult to control in field applications and presents a distinct disadvantage with fluid systems.

These systems contain toxic and hazardous chemicals that require safety protection during installation and disposal of materials. Fluid-applied systems are available in the following derivatives: urethane (single or two-component systems), rubber derivatives (butyl, neoprene, or hypalons), polymeric asphalt, coal tar, or asphalt modified urethane, PVC, and hot applied systems (asphalt).

Urethane

Urethane systems are available in one- or two-component materials. Black coloring is added only to make those people who believe waterproofing is still "black mastic" comfortable with the product. Urethanes are solvent based, requiring substrates to be completely dry to avoid membrane blistering.

These systems have the highest elastomeric capabilities of fluid-applied membranes, averaging 500–750 percent by standardized testing. Urethanes have good resistance to all chemicals likely to be encountered in below-grade conditions, as well as resistance against alkaline conditions of masonry substrates.

Rubber derivatives

Rubber derivative systems are compounds of butyls, neoprenes, or hypalons in a solvent base. Solvents make these materials flammable and toxic. They have excellent elastomeric capability but less than that of urethane membranes.

Rubber systems are resistant to environmental chemicals likely to be encountered below grade. As with most fluid membranes, toxicity requires safety training of mechanics in their use and disposal.

Polymeric asphalt

A chemical polymerization of asphalts improves the generic asphalt material qualities sufficiently to allow their use as a below-grade waterproofing material. Asphaltic compounds do not require drying and curing of a masonry substrate, and some manufacturers allow installation of their asphalt membranes over uncured concrete.

However, asphalt materials are not resistant to chemical attack as are other fluid systems. These membranes have limited life cycling and are used less frequently than other available systems.

Coal tar or asphalt-modified urethane

Coal tar and asphalt-modified urethane systems lessen the cost of the material while still performing effectively. Extenders of asphalt or coal tar limit the elastomeric capabilities and chemical resistance of these membranes.

Coal tar derivatives are especially toxic and present difficulties in installing in confined spaces such as small planters. Coal tar can cause burns and irritations to exposed skin areas. Field mechanics should take necessary precautions to protect themselves from the material's hazards.

Polyvinyl chloride

Solvent-based PVC or plastics are not extensively used in liquid-applied waterproofing applications. These derivatives are more often used as sheet membranes for roofing. Their elastomeric capabilities are less than other fluid systems and have higher material costs. They do offer high resistance to chemical attack for below-grade applications.

Hot-applied fluid systems

Hot-applied systems are improvements over their predecessors of coal tar pitch and felt materials. These systems add rubber derivatives to an asphalt base for improved performance, including crack bridging capabilities and chemical resistance.

Hot systems are heated to approximately 400°F in specialized equipment and applied in thickness up to 180 mil, versus urethane millage of 60 mil (see Fig. 2.6). Asphalt extenders keep costs competitive even at this higher millage. These materials have a considerably extended shelf life compared to solvent-based products, which lose their usefulness in 6 months to 1 year.

Since these materials are hot applied, they can be applied in colder temperatures than solvent-based systems, which cannot be applied in weather under 40°F. Manufacturers often market their products as self-healing membranes, but in below-grade conditions this is a questionable characteristic. Properties of typical fluid-applied systems are summarized in Table 2.3.

Fluid System Application

Substrate preparation is critical for proper installation of fluid-applied systems. See Fig. 2.7 for typical fluid system application. Horizontal concrete surfaces should have a light broom finish for proper bonding. Excessively smooth concrete requires acid etching or sandblasting to roughen the surface for adhesion. Vertical concrete surfaces with plywood form finish are satisfactory, but honeycomb, tie holes, and voids must be patched, with fins and protrusions removed.

Wood surfaces must be free of knotholes or patched before fluid application. Butt joints in plywood decks should be sealed with a compatible sealant followed by a detail coat of membrane. On steel or metal

Figure 2.6 Hot-applied fluid system application. (*Courtesy of Western Group*)

TABLE 2.3 Properties of Fluid-Applied Systems

Advantages	Disadvantages
Excellent elastomeric properties	Application thickness controlled in field
Ease of application	Not applicable over damp or uncured surfaces
Seamless application	Toxic chemical additives

surfaces, including plumbing penetrations, metal must be cleaned and free of corrosion. PVC piping surfaces are roughened by sanding before membrane application.

Curing of concrete surfaces requires a minimum of 7 days, preferably 28 days. On subslabs, shorter cure times are acceptable if concrete passes a mat dryness test. Mat testing is accomplished by tapping visquene to a substrate area. If condensation occurs within 4 hours, concrete is not sufficiently cured or is too wet for applying material.

Blistering will occur if materials are applied to wet substrates since they are nonbreathable coatings. Water curing is the recommended method of curing, but some manufacturers allow sodium silicate curing compounds. Most manufacturers do not require primers over concrete or masonry surfaces; however, metal substrates should be primed.

All cold joints, cracks, and changes in plane should be sealed with sealant followed by a 50–60-mil membrane application, 4-in wide.

Figure 2.7 Fluid system roller application. (*Courtesy of Western Group*)

Cracks over 1/16 in should be sawn out, sealed, then coated. Refer to Fig. 2.8 for these detailing examples.

At wall–floor intersections, a sealant cant approximately ½–1 in high at 45° should be applied followed with a 50-mil detail coat. All projections through a substrate should be similarly detailed. Refer again to Fig. 2.8 for typical installation detailing. At expansion joints and other high movement details, a fiberglass mesh or sheet flashing is

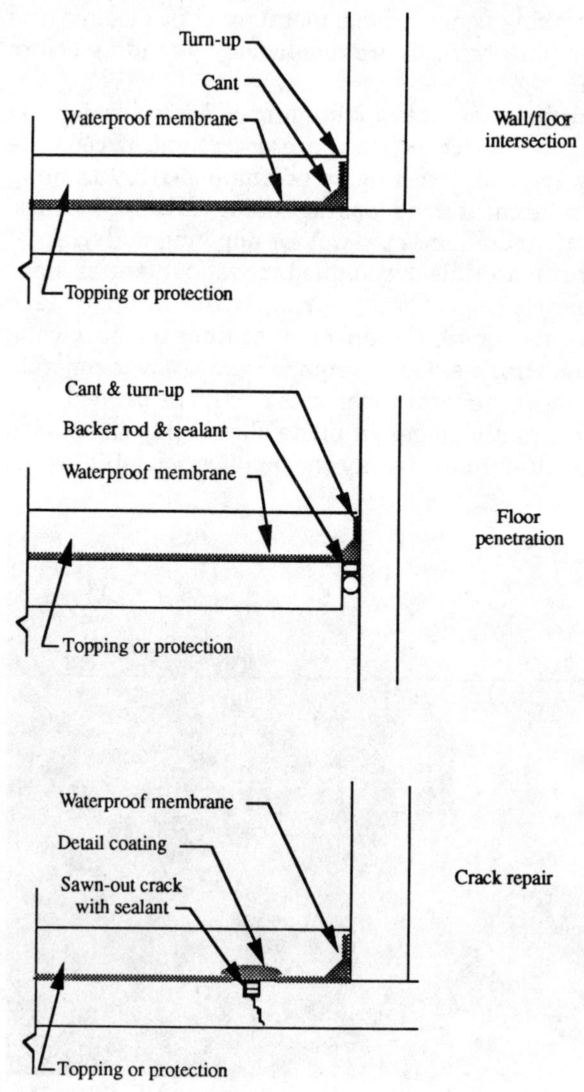

Figure 2.8 Below-grade fluid membrane details.

embedded in the coating material. This allows greater movement capability.

Control coating thickness by using notched squeegees or trowels. If spray equipment is used, take wet millage tests at regular intervals during installation. Application by roller is not recommended. Pinholes in materials occur if a substrate is excessively chalky or dusty, material cures too fast, or material shrinks by improper millage application.

Fluid membranes are supplied in 5- or 55-gal containers. Their toxicity requires proper disposal methods of containers after use. Since these materials rapidly cure when exposed to atmospheric conditions, unopened sealed containers are a necessity.

These materials are not designed for exposed finishes. They will not withstand traffic or ultraviolet weathering. Apply protection surfaces to both horizontal and vertical applications. On vertical surfaces, a ½-in polystyrene material or other lightweight protection system is used. For horizontal installations a ⅛-in asphalt impregnated board is necessary. On curved surfaces, such as tunnel work, 90-lb roll roofing is usually acceptable protection.

Sheet Membrane Systems

Although single-ply systems are frequently used as roof systems, they have not received acceptance in waterproofing applications. Difficult vertical installations, protrusion detailing, and small confined area applications have limited use of single-ply waterproofing materials.

Thermoplastics, vulcanized rubbers, and rubberized asphalts used in waterproofing applications are also used in single-ply roofing applications. However, an experienced roofing mechanic is not necessarily an experienced waterproofer. Although all systems are similar as a generic grouping of waterproofing systems, consider their individual characteristics whenever choosing systems for particular installations.

Sheet membranes have thickness controlled by factory manufacturing. This ensures uniform application thickness throughout an installation. Sheet manufactured systems range in thickness from 20 to 120 mil. Roll goods of materials vary in width from 3 to 10 ft. Larger widths are limited to horizontal applications because they are too heavy and difficult to control for vertical applications.

Unlike liquid systems, sheet system installations involve multiple seams and laps and are not self-flashing at protrusions and changes in plane. This is also true for terminations or transitions into other members of the building envelope.

Applications below grade require protection board during backfill operations and concrete and steel placements. Fins and sharp protrusions in substrates should be removed before application or they will puncture during installation. Materials used in vertical applications should not be left exposed for any length of time before backfilling. Weathering will cause blistering and disbonding if backfill operations must begin immediately after membrane application.

Vertical single-ply applications are more difficult than fluid applications due to the difficulty of handling and seaming materials. Seams are lapped and sealed for complete waterproofing. In small, confined areas such as planter work, vertical installation and transitions to horizontal areas become difficult and extra care must be taken.

Thermoplastics

Thermoplastic sheet-good systems are available in three compositions: PVC, chlorinated polyurethane (CPE), and chlorosulfonated polyethylene (CSPE), which is referred to as hypalon. Materials are manufactured in rolls of varying widths, but difficulty with vertical applications makes smaller widths more manageable.

On horizontal applications, wider roll widths require fewer seams; therefore it is advantageous to use the widest workable widths. All three systems adhere by solvent-based adhesives or heat welding at seams.

PVC membranes are available in thicknesses of 30–60 mil. CPE systems vary by as much as 20–120 mil, and hypalon materials (CSPE) are 30–35 mil. All derivatives have excellent hydrostatic and chemical resistance to below-grade application conditions. PVC membranes are generically brittle materials requiring plasticizers for better elastomeric properties, but elongation of all systems is acceptable for below-grade conditions.

Vulcanized rubbers

Vulcanized rubbers are available in butyl, ethylene propylene diene monomer (EPDM), and neoprene rubber. These materials are vulcanized by the addition of sulfur and heat to achieve better elasticity and durability properties. Membrane thickness for all rubber systems range from 30–60 mil. These materials are nonbreathable and will disbond or blister if negative vapor drive is present.

As with thermoplastic materials, vulcanized rubbers are available in rolls of varying widths. Seam sealing is by a solvent-based adhesive, as heat welding is not applicable. A separate adhesive application to vertical areas is necessary before applying membranes. Vulcanized rubber systems incorporate loosely laid applications for horizontal installations.

Although other derivatives of these materials, such as visquene, are used beneath slabs as dampproofing membranes or vapor barriers, they are not effective if hydrostatic pressure exists. Material installations under slabs on grade by loose laying over compacted fill and sealing joints with adhesive or heat welding are useful in limited waterproofing applications.

This is a difficult installation procedure and usually not specified or recommended. Loosely laid applications do, however, increase the elastomeric capability of the membrane versus fully adhered systems that restrict membrane movement.

Rubberized asphalts

Rubberized asphalt sheet systems originally evolved for use in pipeline protection applications. Sheet goods of rubberized asphalt are available in self-adhering rolls with a polyethylene film attached. Self-adhering membranes adhere to themselves, eliminating the need for a seam adhesive. Sheets are manufactured in varying widths of 3–4 ft and typically 50-ft lengths.

Also available are rubberized asphalt sheets reinforced with glass cloth weave that require compatible asphalt adhesives for adhering to a substrate. Rubber asphalt products require a protection layer to prevent damage during backfill or concrete placement operations.

Self-adhering asphalt membranes include a polyethylene film that acts as an additional layer of protection against water infiltration and weathering. The self-adhering portion is protected with a release paper, which is removed to expose the adhesive for placement. Being virtually self-contained, except for primers, this system is the simplest of all sheet materials to install.

Self-adhering membranes are supplied in 60-mil thick rolls, and accessories include compatible liquid membranes for detailing around protrusions or terminations. Rubberized asphalt systems have excellent elastomeric properties but are not used in above-grade exposed conditions. However, membrane use in sandwich or split-slab construction for above-grade installations is acceptable.

Glass cloth reinforced rubber asphalt sheets, unlike self-adhering systems, require no concrete curing time. Separate adhesive and seam sealers are available. Glass cloth rubber sheets are typically 50 mil thick and require a protection layer for both vertical and horizontal applications. Typical properties of sheet materials are summarized in Table 2.4.

Sheet System Application

Unlike liquid-applied systems, broom finished concrete is not acceptable, as coarse finishes will puncture sheet membranes during application. Concrete must be smoothly finished with no voids, honeycombs,

TABLE 2.4 Sheet Waterproofing Material Properties

Advantages	Disadvantages
Manufacturer-controlled thickness	Vertical applications difficult
Wide rolls for horizontal applications	Seams
Good chemical resistance	Detailing around protrusions difficult

fins, or protrusions. Concrete curing compounds should not contain wax, oils, or pigments. Concrete surfaces must be dried sufficiently to pass a mat test before application.

Wood surfaces must be free of knotholes, gouges, and other irregularities. Butt joints in wood should be sealed with a 4-in-wide membrane detail strip then installed. Masonry substrates should have all mortar joints struck flush. If masonry is rough, a parge coat of cement and sand is required to smooth surfaces.

Metal penetrations should be cleaned, free of corrosion, and primed. Most systems require priming to improve adhesion effectiveness and prevent concrete dust from interfering with adhesion.

All sheet materials should be applied so that seams shed water. This is accomplished by starting at low points and working upward toward higher elevations. With adhesive systems, adhesives should not be allowed to dry before membrane application. Self-adhering systems are applied by removing a starter piece of release paper or polyethylene backing, adhering membrane to substrate.

With all systems, chalk lines should be laid for seam alignment. Seam lap requirements vary from 2 to 4 in. Misaligned strips should be removed and reapplied, with material cut and restarted if alignments are off after initial application. Attempts to correct alignment by pulling on membrane to compensate may cause "fish mouths" or blisters. A typical sheet membrane application is shown in Fig. 2.9.

At changes in plane or direction, manufacturers call for a seam sealant to be applied over seam end laps and membrane terminations. Materials are back rolled at all seams for additional bonding at laps. Any patched areas in the membrane should be rolled to ensure adhesion.

Each manufacturer has specific details for use at protrusions, joints, and change in plane. Typically, one or two additional membrane layers are applied in these areas and sealed with seam sealant or adhesive. Small detailing is sealed with liquid membranes that are compatible and adhere to the sheet material.

Protection systems are installed over membranes before backfilling, placement of reinforcing steel, and concrete placement. Hardboard, ⅛–¼ in thick, made of asphalt impregnated material is used for horizontal applications. Vertical surfaces use polystyrene board, ½ in thick, which

Figure 2.9 Rubberized asphalt sheet membrane application. *(Courtesy of Western Group)*

is lightweight and applied with adhesives to keep it in place during backfill. Sheet systems cannot be left exposed, and backfill should occur immediately after installation.

Hot-Applied Sheet Systems

Hot-applied systems are effectively below-grade roofing systems. They use either coal tar pitch or asphalts with 30-lb roofing felts applied in three to five plies. Waterproofing technology has provided better performance materials and simpler applications, limiting hot systems usage for waterproofing applications.

Hot systems are extremely difficult to apply to vertical surfaces due to the weight of felts. Also, roofing asphalts and coal tars are self-leveling in their molten state, which causes material to flow down walls during application. Safety concerns are multiplied during their use as waterproofing because of difficulties in working with confined areas encountered at below-grade details.

Hot-applied sheet systems have installation and performance characteristics similar to those of roofing applications. These systems are brittle and maintain very poor elastic properties. Extensive equipment and labor costs offset inexpensive material costs. Below-grade areas must be accessible to equipment used for heating materials. If materials are carried over a distance, they begin to cool and cure, providing unacceptable installations. Properties of typical hot-applied sheet systems are summarized in Table 2.5.

TABLE 2.5 Material Properties of Hot-Applied Sheet Systems

Advantages	Disadvantages
Material costs	Safety
Similar to built-up roofing	Difficult vertical installations
Some installations are self-healing	Poor elastomeric properties

Clay Systems

Natural clay systems, commonly referred to as bentonite, are composed primarily of montmorillonite clay. This natural material is used commercially in a wide range of products including toothpaste. Typically, bentonite waterproofing systems contain 85–90 percent of montmorillonite clay and a maximum of 15 percent natural sediments such as volcanic ash.

After being installed in a dry state, clay, when subjected to water, swells and becomes impervious to water. This natural swelling is caused by its molecular structure form of expansive sheets that can expand massively. The amount of swelling and the ability to resist water is directly dependent on grading and clay composition. Clay swells 10–15 percent of its dry volume under maximum wetting. Therefore, it is important to select a system high in montmorillonites and low in other natural sediments.

Bentonite clay is an excellent waterproofing material, but it must be hydrated properly for successful applications. Clay hydration must occur just after installation and backfilling, since the material must be fully hydrated and swelled to become watertight. This hydration and swelling must occur within a confined area after backfill for the waterproofing properties to be effective. Precaution must be taken to ensure the confined space is adequate for clay to swell. If insufficient, materials can raise floor slabs or cause concrete cracking due to the swelling action.

There is no concrete cure time necessary, and minimal substrate preparation is necessary. Of all waterproofing systems, these are the least toxic and harmful to the environment. Clay systems are self-healing, unless materials have worked away from a substrate. Installations are relatively simple, but clay is extremely sensitive to weather during installation. If rain occurs or groundwater levels rise and material is wetted before backfilling, hydration will occur prematurely and waterproofing capability will be lost, since hydration occurred in an unconfined space.

Immediate protection of applications is required, including uses of polyethylene covering to keep materials from water sources before backfill. If installed in below-grade conditions where constant wetting and drying occurs, clay will eventually deteriorate and lose its waterproofing capabilities. These systems should not be installed where free-flowing groundwater occurs, as clay will be washed away from the substrate. Bentonite clays are not particularly resistant to chemicals present in groundwater such as brines, acids, or alkalines.

Bentonite material derivatives are now being added to other waterproofing systems such as thermoplastic sheets and rubberized asphalts. These systems were developed because bulk bentonite spray applications cause problems including thickness control and substrate adhesion. Bentonite systems are currently available in the following forms:

- Bulk
- Fabricated paper panels
- Sheet goods
- Bentonite and rubber combination sheets
- Textile mats

Bulk bentonite

Bulk bentonite is supplied in bulk form and spray applied with an integral adhesive to seal it to a substrate. Applications include direct installations to formwork or lagging before foundation completion in lieu of applications directly to substrates. Materials are applied at quantities of 1–2 lb/ft^2.

Bulk bentonite spray applications provide completely seamless installations. Controls must be provided during application to check that sufficient material is being applied uniformly. Materials should be protected by covering them with polyethylene after installation. Due to possibilities of insufficient thickness during application, manufacturers have developed several clay systems controlling thickness by factory manufacturing, including boards, sheets, and mat systems.

Panel systems

Bentonite clays are packaged in cardboard panels usually 4 ft square, containing 1 lb/ft^2 of bentonite material. Panels are fastened to substrates by nails or adhesives. Upon backfilling, panels deteriorate by anaerobic action, allowing groundwater to cause clay swelling for waterproofing properties.

These systems require time for degradation of cardboard panels before swelling and watertightness occurs. This can allow water to penetrate a structure before swelling occurs. As such, manufacturers have developed systems with polyethylene or butyl backing to provide temporary waterproofing until hydration occurs.

Panel clay systems require the most extensive surface penetration of clay systems. Honeycomb and voids should be filled with clay gels before panel application. Special prepackaged clay is provided for application to changes in plane, and gel material is used at protrusions for detailing.

Several grades of panels are available for specific project installation needs. These include special panels for brine groundwater conditions and reinforced panels for horizontal applications where steel reinforcement work is placed over panels. Panels are lapped onto all sides of adjacent panels using premarked panels that show necessary laps.

Bentonite sheets

Bentonite sheet systems are manufactured by applying bentonite clay at 1 lb/ft^2 to a layer of chlorinated polyethylene. They are packaged in rolls 4 ft wide. The addition of polyethylene adds temporary waterproofing protection during clay hydration. This polyethylene also protects clay material from prematurely hydrating if rain occurs before backfilling and adds chemical-resistant properties to these systems.

Some manufacturers have developed sheet systems for use in above-grade split or sandwich slab construction. However, constant wetting and drying of this system can alter the clay's natural properties, and waterproofing then depends entirely upon the polyethylene sheet.

Bentonite and rubber sheet membranes

Bentonite and rubber sheet membrane systems add clay to a layer of polyethylene, but also compound the bentonite in a butyl rubber composition. Materials are packaged in rolls 3 ft wide that are self-adhering using a release paper backing. They are similar to rubberized asphalt membranes in application and performance characteristics.

These combination sheet systems are used for horizontal applications, typically split-slab construction in parking or plaza deck construction. As with rubberized asphalt systems, accessories must be used around protrusions, terminations, and changes in plane. The polyethylene, butyl rubber, and bentonite each act in combination with the others providing substantial waterproofing properties.

Unlike other clay systems, concrete substrates must be dry and cured before application. Care must be taken in design and construction to allow for adequate space for clay swelling.

Bentonite mats

Bentonite mat systems apply clays at 1 lb/ft^2 to a textile fabric similar to a carpet backing. This combination creates a carpet of bentonite material. The coarseness of the fabric allows immediate hydration of clay after backfilling versus a delayed reaction with cardboard panels.

The textile material is not self-adhering, and adhesives or nailing to vertical substrates is necessary. Protection with a polyethylene sheet after installation is used to prevent premature hydration. This system is particularly effective in horizontal applications where the large rolls eliminate unnecessary seams. This lowers installation costs as well as prevents errors in seaming operations.

Properties of typical clay systems are summarized in Table 2.6.

Clay System Application

Natural clay waterproofing materials require the least preparatory work of all below-grade systems. Concrete substrates are not required to be cured except for rubberized asphalt combination systems. Concrete can be damp during installation but not wet enough to begin clay hydration.

Large voids and honeycombs should be patched before application. Minor irregularities are sealed with clay gels. Most concrete curing agents are acceptable with clay systems. Masonry surfaces should have joints stricken flush.

Bentonite materials combined with butyl rubber require further preparation than other clay systems, including a dry surface, no oil or wax curing compounds, and no contaminants, fins, or other protrusions that will puncture materials.

The variety of bentonite systems available means that applications will vary considerably and have similar procedures to the waterproofing systems they resemble in packaging type (e.g., sheet goods). Bulk clay is applied like fluid membranes. Panels and sheets as sheet-good systems and butyl compound–polyethylene systems are applied virtually identically to rubberized asphalt systems.

TABLE 2.6 Material Properties of Clay Systems

Advantages	Disadvantages
Self-healing characteristics	Clay subject to hydration before backfilling
Ease of application	Not resistant to chemical in soil
Range of systems and packaging	Must be applied in confined conditions for proper swelling conditions

With bulk systems, proper material thickness application is critical as it is with fluid-applied systems. Bulk systems are sprayed or troweled applied at 1–2 lb/ft^2 of substrate.

Panel and mat systems are applied to vertical substrates by nailing. Horizontal applications require lapping only. These systems require material to be lapped 2 in on all sides. Cants of bentonite material are installed at changes in plane, much the same way as cementitious or sheet-applied systems. Bentonite sheet materials are applied with seams shedding water by starting applications at low points.

Outside corners or turns receive an additional strip of material usually 1 ft wide for additional reinforcement. Chalk lines should be used to keep vertical applications straight and to prevent fish mouthing of materials. All end laps, protrusions, and terminations should be sealed with the clay mastic.

Vapor Barriers

Vapor barriers are not suitable for waterproofing applications. As their name implies, they prevent transmission of water vapor through a substrate in contact with the soil. Typically used at slabs-on-grade conditions, they also are used in limited vertical applications.

Vapor barriers are sometimes used in conjunction with other waterproofing systems, where select areas of the building envelope are not subject to actual water penetration. Vapor barriers are discussed only to present their differences and unsuitability for envelope waterproofing.

As previously discussed, soils have characteristic capillary action that allows the upward movement or migration of water vapor through the soil. Beginning as water and saturating the soil immediately adjacent to the water source, the capillary action ends as water vapor in the upper capillary capability limits of the soil.

Vapor barriers prevent upward capillary migration of vapor through soils from penetrating pores of concrete slabs. Without such protection, delamination of flooring materials, damage to structural components, paint peeling, mildew formation, and increased humidity in finished areas will occur.

Vapor barriers are produced in PVC, combinations of reinforced waterproof paper with a polyvinyl coating, or polyethylene sheets (commonly referred to as visquene). Polyethylene sheets are available in both clear and black colors in thickness ranging from 5 to 10 mil. PVC materials are available in thickness ranging from 10 to 60 mil. Typical properties of vapor barriers are summarized in Table 2.7.

Vapor barriers are rolled or spread out over prepared and compacted soil with joints lapped 6 in. Vapor barriers can be carried under, up, and over foundations to tie horizontal floor applications into vertical

TABLE 2.7 Material Properties of Vapor Barriers

Advantages	Disadvantages
Ease of horizontal applications	Noneffective waterproofing materials
Prevent moisture transmission	Seams
No subslab required	Difficult vertical installations

application over walls. This is necessary to maintain the integrity of a building envelope.

Mastics are typically available from manufacturers for adhering materials to vertical substrates. In clay soil, where capillary action is excessive, laps should be sealed with a mastic for additional protection. Proper foundation drainage systems should be installed, as with all waterproofing systems.

Vapor barriers are installed directly over soil, which is not possible with most waterproofing systems. Protection layers or boards are not used to protect the barrier during reinforcement application or concrete placement.

Summary

Systems available for below-grade waterproofing are numerous and present sufficient choices for ensuring the integrity of below-grade envelopes. Project conditions to review when choosing an appropriate below-grade waterproofing system include

- Soil conditions; rock or clay soils can harm waterproofing systems during backfill.
- Chemical contamination, especially salts, acids, and alkalines.
- Freeze–thaw cycling and envelope portions below frost line.
- Expected movement, including settlement and differential.
- Concrete cold joints to see if they are treatable for the system selected.
- Positive or negative system to see which is better for job site conditions.
- Large vertical applications, which are difficult with certain systems.
- Difficult termination and transition detailing, which prevents use of many systems.
- Length of exposure of installed system due to project conditions before backfilling.
- Safety concerns at project.

- Dewatering requirements during construction.
- Concrete cure time available before backfill or other construction must commence.
- Adjacent envelope systems that the waterproofing system must be comparable with or not damage.
- Scheduling requirements.
- Access for repairs after construction is complete.

Although not a substitute for referring to a specific manufacturer information on a specific material, Table 2.8 is a summary and comparison of major below-grade waterproofing systems. One system or material may not be sufficient for all situations encountered on a particular project. Once below-grade materials are chosen, they must be detailed into above-grade envelope materials and systems. This detailing is critical to the success of the entire building envelope and is discussed further in Chap. 8.

TABLE 2.8 Summary Properties of Below-Grade Materials

Property	Cementitious	Fluid applied	Sheet goods	Clay system
Elongation	None	Excellent	Good	Good
Chemical and weathering resistance	Good	Fair to Good	Good	Fair
Difficulty of installation	Moderate	Simple	Difficult	Simple
Thickness	$\frac{1}{8}$–$\frac{7}{16}$ in	60 mil average	20–60 mil	$\frac{1}{4}$–$\frac{1}{2}$ in
Horizontal subslab required	No	Yes	Yes	No
Positive or negative system	Both	Positive	Positive	Positive
Areas requiring inspections	Coves and cants at changes in plane; control joint detailing	Millage, especially at turnups; detailing and priming at penetrations	Laps and seams; penetration detailing; transition details	Laps, penetration detailing, changes in plane
Repairs	Simple	Simple	Moderate to difficult	Moderate
Protection required	No	Yes	Yes	No

Chapter

3

Above-Grade Waterproofing

Waterproofing of surfaces above grade is the prevention of water intrusion into exposed elements of a structure or its components. Above-grade materials are not subject to hydrostatic pressure but are exposed to detrimental weathering effects such as ultraviolet light.

Water that penetrates above-grade envelopes does so in five distinct methods:

- Natural gravity forces
- Capillary action
- Surface tension
- Air pressure differential
- Wind loads

The force of water entering by gravity is greatest on horizontal or slightly inclined envelope portions. Those areas subject to ponding or standing water must be adequately sloped to provide drainage away from envelope surfaces.

Capillary action is the natural upward wicking motion that can draw water from ground sources up into above-grade envelope areas. Likewise, walls resting on exposed horizontal portions of an envelope (e.g., balcony decks) can be affected by capillary action of any ponding or standing water on these decks.

The molecular surface tension of water allows it to adhere to and travel along the underside of envelope portions such as joints. This water can be drawn into the building by gravity or unequal air pressures.

If air pressures are lower inside a structure than on exterior areas, water can be literally sucked into a building. Wind loading during heavy rainstorms can force water into interior areas if an envelope is not structurally resistant to this loading. For example, curtain walls and glass can actually bend and flex away from gaskets and sealant joints causing direct access for water.

The above-grade envelope must be resistant to all these natural water forces to be completely watertight. Waterproofing the building envelope can be accomplished by the facade material itself (brick, glass, curtain wall) or by applying waterproof materials to these substrates. Channeling water that passes through substrates back out to the exterior using flashing, weeps, and dampproofing is another method. Most envelopes include combinations of all these methods.

Older construction techniques often included masonry construction with exterior load bearing walls up to 3 ft thick. This type of envelope required virtually no attention to waterproofing or weathering due to the shear impregnability of the masonry wall.

Today, however, it is not uncommon for high-rise structures to have an envelope skin thickness of ⅛ in. Such newer construction techniques have developed from the need for lighter weight systems to allow for simpler structural requirements and lower building costs. These systems in turn create problems in maintaining an effective weatherproof envelope.

Waterproof building surfaces are required at vertical portions as well as horizontal applications such as balconies and pedestrian plaza areas. Roofing is only a part of necessary above-grade waterproofing systems that must be carefully tied into other building envelope components.

Today roofing systems take many different forms of design and detailing. Plaza decks or balcony areas covering enclosed spaces and parking garage floors covering an occupied space all constitute individual parts of a total roofing system. Buildings can have exposed roofs as well as unexposed membranes acting as roofing and waterproofing systems for preventing water infiltration into occupied areas.

Differences from Below-Grade Systems

Most above-grade materials are breathable in that they allow for negative vapor transmission. This is similar to human skin; it is waterproof, allowing you to swim and bathe but also to perspire, which is negative moisture transmission. Most below-grade materials will not allow negative transmission and, if present, cause the material to blister or become unbonded.

Breathable coatings are necessary on all above-grade wall surfaces to allow moisture condensation from interior surfaces to pass through wall structures to the exterior. The sun causes this natural effect by drawing vapors to the exterior. Pressure differentials that might exist between exterior and interior areas create this same condition.

Vapor barrier (nonbreathable) products installed above grade cause spalling during freeze–thaw cycles. Vapor pressure buildup behind a nonbreathable coating will also cause the coating to disbond from substrates. This effect is similar to window or glass areas that are vapor barriers and cause formation of condensation on one side that cannot pass to exterior areas.

Similarly, condensation passes through porous wall areas back out to the exterior when a breathable coating is used but condensates on the back of nonbreathable coatings. This buildup of moisture, if not allowed to escape, will deteriorate structural reinforcing steel and other internal wall components.

Below-grade products are neither ultraviolet resistant nor capable of withstanding thermal movement experienced in above-grade structures. Whereas below-grade materials are not subject to wear, above-grade materials can be exposed to wear such as foot traffic. Below-grade products withstand hydrostatic pressure, whereas above-grade materials do not.

Since many waterproofing materials are not aesthetically acceptable to architects or engineers, some trade-off of complete watertightness versus aesthetics is used or specified. For instance, masonry structures using common face brick are not completely waterproof due to water infiltration at mortar joints. Rather than change the aesthetics of brick by applying a waterproof coating, the designer chooses a dampproofing and flashing system. This dampproofing system diverts water that enters through the brick wall back out to the exterior. Application of a clear water repellent will also reduce water penetration through the brick and mortar joints. Such sealers also protect brick from freeze–thaw and other weathering cycles.

Thus, waterproofing exposed vertical and horizontal building components can include a combination of installations and methods that together compose a building envelope. This is especially true of buildings that use a variety of composite finishes for exterior surfacing such as brick, precast, and curtain wall systems. With such designs, a combination of several waterproofing methods must be used. Although each might act independently, as a whole they must act cohesively to prevent water from entering a structure. Sealants, wall flashings, weeps, dampproofing, wall coatings, deck coatings, and the natural weathertightness of architectural finishes themselves must act together to prevent water intrusion.

This chapter will cover vertical waterproofing materials, including clear water repellents, elastomeric coatings, cementitious coatings, and related patching materials. It will also review horizontal waterproofing materials, including deck coatings, sandwich slab membranes, and roofing.

Vertical Applications

Several systems are available for weatherproofing vertical wall envelope applications. Clear sealers are useful when substrate aesthetics are important. These sealers are typically applied over precast architectural concrete, exposed aggregate, natural stone, brick, or masonry.

It is important to note that clear sealers are not completely waterproof; they merely slow down the rate of water absorption into a substrate, in some situations as much as 98 percent. However, wind-driven rain and excessive amounts of water will cause eventual leakage through any clear sealer system. This requires flashings, dampproofing, sealants, and other systems to be used in conjunction with sealers to ensure drainage of water entering through primary envelope barriers.

This situation is similar to wearing a canvas-type raincoat. During light rain, water runs off; but should the canvas become saturated, water passes directly through the coat. Clear sealers as such are defined as water repellents, in that they shed water flow but are not impervious to water saturation or a head of water pressure.

Elastomeric coatings are high solid content paints that produce high millage coatings when applied to substrates. These coatings are waterproof within normal limitations of movement and proper application. Elastomeric coatings completely cover and eliminate any natural substrate aesthetics. They can, however, add a texture of their own to an envelope system, depending on the amount of sand, if any, in the coating.

To waterproof adequately with an elastomeric coating, details must be addressed, including patching cracks or spalls in substrates, allowing for thermal movement, and installation of flashings where necessary.

Cementitious coatings are available for application to vertical masonry substrates, which also cover substrates completely. The major limitation of cementitious above-grade product use is similar to its below-grade limitation. The products do not allow for any substrate movement or they will crack and allow water infiltration. Therefore, proper attention to details is imperative when using cementitious materials. Installing sealant joints for movement and crack preparation must be completed before cementitious coating application.

With all vertical applications, there are patching materials used to ensure water tightness of the coating applied. These products range from brushable-grade sealants for small cracks, to high-strength, quick set cementitious patching compounds for repairing spalled substrate areas.

Horizontal Applications

Several types of systems and products are available for horizontal above-grade applications, such as parking garages and plaza decks. Surface coatings, which apply directly to exposed surfaces of horizontal substrates, are available in clear siloxane types or solid coatings of urethane or epoxy. Clear horizontal sealers, as with vertical applications, do not change existing substrate aesthetics to which they are applied. They are, however, not in themselves completely waterproof but only water resistant.

Clear coatings are often specified for applications to prevent chloride ion penetration into concrete substrates from such materials as road salts. These pollutants attack reinforcing steel in concrete substrates and cause spalling and structural deterioration.

Urethane, epoxy, or acrylic coatings change the aesthetics of a substrate but have elastomeric properties that allow bridging of minor cracking or substrate movement. Typically, these coatings have a "wearing coat" that contains silicon sand or carbide, which allows vehicle or foot traffic while protecting the waterproof base coat.

Subjecting coatings to foot or vehicular wear requires maintenance at regular frequency and completion of necessary repairs. The frequency and repairs are dependent on the type and quantity of traffic occurring over the envelope coating.

As with vertical materials, attention to detailing is necessary to ensure watertightness. Expansion or control joints must be properly sealed, cracks or spalls in the concrete must be repaired before application, and allowances for drainage must be created.

Several types of waterproof membranes are available for covered decks such as sandwich slab construction or tile-topped decks. These membranes are similar to those used in below-grade applications, including liquid-applied and sheet-good membranes. Such applications are also used as modified roofing systems.

Above-Grade Exposure Problems

All above-grade waterproof systems are vulnerable to a host of detrimental conditions due to their exposure to weathering elements and substrate performance under these conditions. Exposure of the entire above-grade building envelope requires resistance from many severe effects, including

- Ultraviolet weathering
- Wind loading
- Structural loading due to snow or water
- Freeze–thaw cycles
- Thermal movement
- Differential movement
- Mildew and algae attack
- Chemical and pollution attack from chloride ions, sulfates, nitrates, and carbon dioxide

Chemical and pollution attack is becoming increasingly more frequent and difficult to contend with. Chloride ions (salts) are extremely corrosive to the reinforcing steel present in all structures, whether it is structural steel, reinforcing steel, or building components such as shelf angles.

Even if steel is protected by encasement in concrete or is covered with a brick facade, water that penetrates these substrates carries chloride ions that attack the steel. Once steel begins to corrode, it increases greatly in size, causing spalling of adjacent materials and structural cracking of substrates.

All geographic areas are subject to chloride ion exposure. In coastal areas, salt spray is concentrated and spread by wind conditions; in northern climates, road salts are used during winter months. Both increase chloride quantities available for corrosive effects on envelope components.

Acid rain now affects all regions of the world. When sulfates and nitrates present in the atmosphere are mixed with water, they create sulfuric and nitric acids (acid rain), which affect all building envelope components. Acids attack the calcium compounds of concrete and masonry surfaces causing substrate deterioration. They also affect exposed metals on a structure, such as flashing, shelf angles, and lintel beams.

Within masonry or concrete substrates, a process of destructive weathering called carbonation occurs to unprotected, unwaterproofed surfaces. Carbonation is the deterioration of cementitious compounds found in masonry substrates when exposed to the atmospheric pollutant carbon dioxide (automobile exhaust).

Carbon dioxide mixes with water to form carbonic acid, which then penetrates a masonry or concrete substrate. This acid begins deteriorating cementitious compounds that form part of a substrate.

Carbonic acid also causes corrosion of embedded reinforcing steel such as shelf angles by changing the substrate alkalinity that surrounds this steel. Reinforcing steel, which is normally protected by the high alkalinity of concrete, begins to corrode when carbonic acid change lowers alkalinity while also deteriorating the cementitious materials.

Roofing systems will deteriorate because of algae attack. Waterproof coatings become brittle and fail due to ultraviolet weathering. Thermal movement will split or cause cracks in a building envelope. This requires that any waterproof material or component of the building envelope be resistant to all these elements, thus ensuring their effectiveness and, in turn, protecting a building during its life cycling.

Finally, an envelope is also subject to building movement, both during and after construction. Building envelope components must withstand this movement; otherwise, designs must include allowances for movement or cracking within the waterproofing material.

Cracking of waterproofing systems occur because of structural settlement, structural loading, vibration, shrinkage of materials, thermal movement, and differential movement. To ensure successful life cycling of a building envelope, allowances for movement must be made, including expansion and control joints, or materials must be chosen that can withstand expected movement.

All these exposure problems must be considered when choosing a system for waterproofing above-grade envelope portions. Above-grade waterproofing systems include the following horizontal and vertical applications:

- Vertical
  - Clear repellents
  - Cementitious coatings
  - Elastomeric coatings
- Horizontal
  - Deck coatings
  - Clear deck sealers
  - Protected membranes

Clear Repellents

Although clear sealers do not fit the definition of a true waterproofing system, they do add water repellency to substrates where solid coatings as an architectural finish are not acceptable. Clear sealers are applied on masonry or concrete finishes when a repellent that does not change substrate aesthetics is required. Clear sealers are also specified for use on natural stone substrates such as limestone. Water repellents

TABLE 3.1 Repellent Types and Compositions

Penetrating sealers	Film-forming sealers
Siloxanes	Acrylics
Silanes	Silicones
Hydrophobic silica	Aliphatic urethane
Hydrocarbon plastic	Aromatic urethane
Epoxy-modified siloxane	Silicone resin
Silane–silicic combination	Methyl methacrylate
Siloxane–acrylic combination	Modified stearate

prevent chloride ion penetration into a substrate and prevent damage from the freeze–thaw cycles.

There is some disagreement over the use of sealers in historic restoration. Some prefer stone and masonry envelope components to be left natural, repelling or absorbing water and aging naturally. This is more practical in older structures that have massive exterior wall substrates than in modern buildings. Today exterior envelopes are as thin as ⅛ in, requiring additional protection such as clear sealers.

The problem with clear sealers is not in deciding when they are necessary but in choosing a proper material for specific conditions. Clear repellents are available in a multitude of compositions, including penetrates and film-forming materials. They vary in percentage solids content and are available in tint or stain bases to add uniformity to the substrate color.

The multitude of materials available requires careful consideration of all available products to select the material appropriate for a particular situation. Repellents are available in compositions and combinations shown in Table 3.1. Sealers are further classified into penetrating and film-forming sealers.

Clear sealers will not bridge cracks in the substrate, and this presents a major disadvantage in using these materials as envelope components. Should cracks be properly prepared in a substrate before application, effective water repellency is achievable. Should further cracking occur, due to continued movement, however, a substrate will lose its watertightness. Properly designed and installed crack control procedures, such as control joints and expansion joints, alleviate cracking problems.

Film-forming sealers

Film-forming, or surface, sealers have a viscosity sufficient to remain primarily on top of a substrate surface. Penetrating sealers have suffi-

TABLE 3.2 Film-Forming Sealer Properties

Advantages	Disadvantages
High solids content; able to fill minor cracks in substrate	Not effective in weathering
Opaque stains available to cover repair work in substrate	Not resistant to abrasive wear
Applicable to exposed aggregate finishes and wood substrates	Film adhesion dependent on substrate cleanliness and preparation

ciently low viscosity of the vehicle (binder and solvent) to penetrate into masonry substrate pores. The resin molecule sizes of a sealer determine the average depth of penetration into a substrate.

Effectiveness of film-forming and penetrating sealers is based upon the percentage of solids in the material. High solid acrylics will form better films on substrates by filling open pores and fissures and repelling a greater percentage of water. Higher solids content materials are necessary when used with very porous substrates; however, these materials may darken or impart a glossy, high sheen appearance to a substrate.

Painting or staining over penetrating sealers is not recommended as it defeats the purpose of the material. With film-forming materials, if more than a stain is required, it may be desirable to use an elastomeric coating to achieve the desired watertightness and color.

Most film-forming materials and penetrates are available in semi-transparent or opaque formulations. If it is desired to add color or a uniform coloring to a substrate that may contain color irregularities (such as tilt-up or poured-in-placed concrete), these sealers offer effective solutions. (See Table 3.2.)

Penetrating sealers

Penetrating sealers are used on absorptive substrates such as masonry block, brick, concrete, and porous stone. Some penetrating sealers are manufactured to react chemically with these substrates, forming a chemical bond that repels water. Penetrating sealers are not used over substrates such as wood, glazed terra cotta, previously painted surfaces, and exposed aggregate finishes.

On these substrates, film-forming clear sealers are recommended (which are also used on masonry and concrete substrates). These materials form a film on the surface that acts as a water-repellent barrier. This makes a film material more susceptible to erosion due to ultraviolet weathering and abrasive wear such as foot traffic.

TABLE 3.3 Penetrating Sealer Properties

Advantages	Disadvantages
Resistant to ultraviolet weathering	Can damage adjacent substrates, especially glass and aluminum
Effective for abrasive wear areas	Causes damage to plants and shrubs
Excellent permeability ratings	Not effective on wood or hard finish materials such as glazed tile

Penetrating sealers are breathable coatings in that they allow water vapor trapped in a substrate to escape through the coating to the exterior. Film-forming sealers' vapor transmission (perm rating) characteristics are dependent on their solids content. Vapor transmission or perm ratings are available from manufacturers. Permeability is an especially important characteristic for masonry installed at grade line. Should an impermeable coating be applied here, moisture absorbed into masonry by capillary action from ground sources will damage the substrates, including surface spalling.

Many sealers fail due to a lack of resistance to alkaline conditions found in concrete and masonry building materials. Most building substrates are high in alkalinity, which causes a high degree of failure with poor alkaline-resistant sealers.

Penetrating materials usually have lower coverage rates and higher per-gallon costs than film materials. Penetrating sealers, however, require only a one-coat application versus two for film-forming materials, reducing labor costs.

Penetrating and film-forming materials are recognized as effective means of preventing substrate deterioration due to acid rain effects. They prevent deterioration from air and water pollutants and from dirt and other contaminants by not allowing these pollutants to be absorbed into a substrate. (See Table 3.3.)

Sealer testing

Several specific tests should be considered in choosing clear sealers. Testing most often referred to is the National Cooperative Highway Research Program (NCHRP). Although often used for testing horizontal applications, it remains an effective test for vertical sealers as well. NCHRP test 244, series 11, measures the weight gain of a substrate by measuring water absorption into a test cube submerged after treatment with a selected water repellent. To be useful a sealer should limit weight gain to less than 20 percent of original weight and preferably less than 15 percent. Test results are also referred to as a reduction in water absorption from the control (untreated) cube. These limits should be a 75 to 100 percent reduction, preferably above 85 percent.

Testing by ASTM includes ASTM E-514, Water Permeability of Masonry, ASTM C-67, Water Repellents Test, and ASTM C-642, Water Absorption Test. Also, federal testing by test SS-W-11OC includes water absorption testing.

Any material chosen for use as a clear sealer should be tested by one of these methods to determine water absorption or repellency. Effective water repellency should be above 85 percent, and water absorption should be less than 25 percent, preferably 20 to 15 percent. Tables 3.2 and 3.3 summarize the advantages and disadvantages of film-forming versus penetrating sealer properties. Testing of materials and envelope components is covered in detail in Chap. 10.

Acrylics

Acrylics and their derivatives, including methyl methacrylates, are film-forming repellents. Acrylics are formulated from copolymers of acrylic or methocrylic acids. Their penetration into substrates is minimal, and they are, therefore, considered film-forming sealers. Acrylic derivatives differ by manufacturer, each having its own proprietary formulations.

Acrylics are available in both water- and solvent-based derivatives. They are frequently used when penetrating sealers are not acceptable for substrates such as exposed aggregate panels, wood, and dense tile. They are also specified for extremely porous surfaces where a film buildup is desirable for water repellency.

Acrylics do not react chemically with a substrate and form a barrier by filming over surfaces as does paint. Solids content of acrylics vary from 5 to 48 percent. The higher a solids content the greater the amount of sheen imparted to a substrate. High solids materials are sometimes used or specified to add a high gloss or glazed appearance to cementitious finish materials such as plaster. Methyl methacrylates are available in 5 to 25 percent solids content.

Most manufacturers require two-coat applications of acrylic materials for proper coverage and uniformity. Coverage rates vary depending on the substrate and its porosity, with first coats applied at 100–250 ft^2/gal. Second coats are applied 150–350 ft^2/gal. Acrylics should not be applied over wet substrates as solvent-based materials may turn white if applied under these conditions. They also cannot be applied in freezing temperatures or over a frozen substrate.

Higher solids content acrylics have the capability of being applied in sufficient millage to fill minor cracks or fissures in a substrate. How-

TABLE 3.4 Acrylic Water Repellent Properties

Advantages	Disadvantages
High solids materials can fill minor substrate cracks and fissures	Not good weathering resistance
Stain colors available; compatible with patching materials	Can pick up dirt particles during cure stage
Breathable coating, allows vapor transmission	Poor crack bridging capabilities

ever, no acrylic is capable of withstanding movement from thermal or structural conditions. Acrylic sealers have excellent adhesion when applied to properly prepared and cleaned substrates. Their application resists the formation of mildew, dirt buildup, and salt and atmospheric pollutants.

Acrylics are available in transparent and opaque stains. This coloring enables hiding or blending of repairs to substrates with compatible products such as acrylic sealants and patching compounds. Stain products maintain existing substrate textures and do not oxidize or peel as paint might.

Acrylics are compatible with all masonry substrates including limestone, wood, aggregate panels, and stucco that has not previously been sealed or painted. Acrylic sealers are not effective on very porous surfaces such as lightweight concrete block. The surface of this block contains thousands of tiny gaps or holes filled with trapped air. The acrylic coatings cannot displace this trapped air and are ineffective sealers over such substrates. (See Table 3.4.)

Silicones

Silicone-based water repellents are manufactured by mixing silicone solids (resins) into a solvent carrier. Most manufacturers base their formulations on a 5 percent solids mixture in conformance with the requirements of federal specification SS-W-11OC.

Although most silicone water repellents are advertised as penetrating, they function as film-forming sealers. Being a solvent base allows the solid resin silicone to penetrate the surface of a substrate but not to depths that siloxanes or quartz carbide sealers penetrate. The silicone solids are deposited onto the capillary pores of a substrate, effectively forming a film of solids that repel water.

All silicone water repellents are produced from the same basic raw material, silane. Manufacturers are able to produce a wide range of repellents by combining or reacting different compounds with this base silane material. These combinations result in a host of silicone-based repellents, including generic types of siliconates, silicone resins, silicones, and siloxanes. The major difference in each of these derivatives is its molecular size.

Regardless of derivative type, molecular size, or compound structure, all silicone-based repellents repel water in the same way. By penetrating substrates, they react chemically with atmospheric moisture, by evaporation of solvents, or by reaction with atmospheric carbon dioxide to form silicone resins that repel water.

Only molecular sizes of the final silicone resin are different. Silicone-based products require that silica be present in a substrate for the proper chemical actions to take place. Therefore, these products do not work on substrates such as wood, metal, or natural stone.

A major disadvantage of silicone water repellents is their poor weathering resistance. Ultraviolet intense climates can quickly deteriorate these materials and cause a loss of their water repellency. Silicone repellents are not designed for horizontal applications, as they do not resist abrasive wearing.

Silicone repellents are inappropriate for marble or limestone substrates, which discolor if these sealer materials are applied. Discoloring can also occur on other substrates such as precast concrete panels. Therefore, any substrate should be checked for staining by a test application with the proposed silicone repellent.

Lower solid concentration materials of 1 to 3 percent solids are available to treat substrates subject to staining with silicone. These formulations should be used on dense surface materials such as granite to allow proper silicone penetration. Special mixes are manufactured for use on limestone but also should be tested before actual application. Silicones can yellow after application, aging, or weathering.

As with most sealers, substrates will turn white or discolor if applied during wet conditions. Silicones do not have the capabilities to span or bridge cracking in a substrate. Very porous materials, such as lightweight or split-face concrete blocks, are not acceptable substrates for silicone sealer application. Adjacent surfaces such as windows and veg-

TABLE 3.5 Silicone Water Repellent Properties

Advantages	Disadvantages
Breathable coating, allows vapor transmission	Poor ultraviolet resistance
Easy application	Can stain or yellow a substrate such as limestone
Cost	Contamination of substrate prohibits other materials application over silicone

TABLE 3.6 Urethane Water Repellent Properties

Advantages	Disadvantages
Applications over wood and metal substrates	Poor vapor transmission
Horizontal applications also	Blisters that occur if applied over wet substrates
Chemical-, acid-, and solvent-resistant applications	Higher material cost

etation should be protected from overspray during application. (See Table 3.5.)

Urethanes

Urethane repellents, aliphatic or aromatic, are derivatives of carbonic acid, a colorless crystalline compound. Clear urethane sealers are typically used for horizontal applications but are also used on vertical surfaces. With a high solids content averaging 40 percent, they have some ability to fill and span nonmoving cracks and fissures up to 1/16 in wide. High solids materials such as urethane sealers have low perm ratings that cause coating blistering if any moisture or vapor drive occurs in the substrate.

Urethane sealers are film-forming materials that impart a high gloss to substrates, and they are nonyellowing materials. They are applicable to most substrates including wood and metal, but adhesive tests should be made before each application. Concrete curing agents can create adhesion failures if the surface is not prepared by sandblasting or acid etching.

Urethane sealers can also be applied over other compatible coatings, such as urethane paints, for additional weather protection. They are resistant to many chemicals, acids, and solvents and are used on stadium structures for both horizontal and vertical seating sections. The cost of urethane materials has limited their use as sealers. (See Table 3.6.)

Silanes

Silanes contain the smallest molecular structures of all silicone-based materials. The small molecular structure of the silane allows the deepest penetration into substrates. Silanes, like siloxanes, must have silica present in substrates for the chemical action to take place that provides water repellency. These materials cannot be used on sub-

TABLE 3.7 Silane Water-Repellent Properties

Advantages	Disadvantages
Deepest penetration capabilities of all silicone-based products	High evaporation rate during application
Forms chemical bond with substrate with good permeability rating	Dry substrates must be wetted to ensure chemical reactions before evaporation
Good weathering characteristics	High cost of material

strates such as wood, metal, or limestone that have no silica present for chemical reaction.

Of all the silicone-based materials, silanes require the most difficult application procedures. Substrates must have sufficient alkalinity in addition to the presence of moisture to produce the required chemical reaction to form required silicone resins. Silanes have high volatility that causes much of the silane material to evaporate before the chemical reaction forms the silicon resins. This evaporation causes a high silane concentration, as much as 40 percent, to be lost through evaporation.

Should a substrate become wet too quickly after application, the silane is washed out from the substrate prohibiting proper water repellency capabilities. If used during extremely dry weather, after application substrates are wetted to promote the chemical reaction necessary. The wetting must be done before all the silane evaporates.

As with other silicone-based products, silanes applied properly form a chemical bond with a substrate. Silanes have a high repellency rating when tested in accordance with ASTM C-67, with some products achieving repellency over 99 percent. As with urethane sealers, their high cost limits their usage. (See Table 3.7.)

Siloxanes

Siloxanes are produced from the CL-silane material, as are other silicone masonry water repellents. Siloxanes are used more frequently than other clear silicones, especially for horizontal applications. Siloxanes are manufactured in two types, oligomerous (short chain of molecular structure) and polymeric (longer chain of molecular structure) alkylalkoxysiloxanes.

Most siloxanes produced now are oligomerous. Polymeric products tend to remain wet or tacky on the surface, attracting dirt and pollutants. Also, polymeric siloxanes have poor alkali resistance, and alkalis are common in masonry products for which they are intended. Oligomerous siloxanes are highly resistant to alkaline attack, and therefore

TABLE 3.8 Siloxanes Water-Repellent Properties

Advantages	Disavantages
Not susceptible to alkali degradation	Not applicable on natural stone substrates
Bonds chemically with substrates with high permeability rating	Can damage adjacent substrates and vegetation
High repellency rating and excellent penetration depth	Cost

can be used successfully on high alkaline substrates such as cement-rich mortar.

Siloxanes react with moisture, as do silanes, to form the silicone resin that acts as the water-repellent substance. Upon penetration of a siloxane into a substrate it forms a chemical bond with the substrate. The advantage of siloxanes over silanes is that their chemical structure does not promote a high evaporation rate.

The percentage of siloxane solids used is substantially less (usually less than 10 percent for vertical applications), thereby reducing costs. Chemical reaction time is achieved faster with siloxanes, which eliminates a need for wetting after installation. Repellency is usually achieved within 5 hours with a siloxane.

Siloxane formulations are now available that form silicone resins without the catalyst—alkalinity—required. Chemical reactions with siloxanes take place even with a neutral substrate as long as moisture, in the form of humidity, is present.

These materials are suitable for application to damp masonry surfaces without the masonry turning white, which might occur with other materials. Testing of all substrates should be completed before full application to ensure compatibility and effectiveness of the sealer.

Siloxanes do not change porosity or permeability characteristics of a substrate. This allows moisture to escape without damaging building materials or the repellent. Since siloxanes are not subject to high evaporation rates, they can be applied successfully by high-pressure sprays for increased labor productivity.

Siloxanes, as other silicone-based products, may not be used with certain natural stones such as limestone. They are also not applicable to gypsum products or plaster. Siloxanes should not be applied over painted surfaces, and if surfaces are to be painted after treatment they should be first tested for compatibility. (See Table 3.8.)

Diffused quartz carbide

Hydrophobia is defined as a dread of water. Hydrophobic silicas, commonly referred to as diffused quartz carbides, are relatively recent de-

TABLE 3.9 Diffused Quartz Carbide Water-Repellent Properties

Advantages	Disadvantages
Good repellency rates	Not applicable over natural stone, gypsum, or asphaltic substrates
Bonds chemically to substrate	Cannot be applied over damp substrates
High permeability rates	Exposed vegetation and adjacent substrates must be protected from over spray

velopments in water-repellent technology. The silica, diffused quartz carbide, is suspended in a petroleum base of hydrocarbon solvent. Upon application, the sealer penetrates a substrate. The solvent evaporates and the silica bonds chemically with the substrate to create water-repellent properties.

As with all silicone-based materials, diffused quartz carbide materials are used only with cementitious products such as concrete, masonry, or stucco. They are not recommended for use with natural stone products that discolor or prevent the necessary chemical reaction from taking place.

Silica water repellents do not interfere with the natural porosity or permeability of a substrate. However, unlike siloxanes, quartz carbide products are not applied to damp or partially damp surfaces. If applied under wet conditions, a substrate is likely to whiten or have streaks appear.

Hydrophobic sealers are not intended for use on gypsum, asphalt, or previously painted surfaces. In addition, if it is necessary to paint over a silicone-treated surface, a test sample should be completed to ensure compatibility. The diffused quartz products are available in stain base if substrate coloring is desired.

High evaporation rates of the solvents used in diffused quartz carbide materials prevent use of most high-pressure spray application equipment. Low-pressure spray application is recommended (e.g., Hudson or garden-type sprayer). Manufacturers usually recommend that a mist or fog coat first be applied to a substrate to break or release surface tension. This allows greater material penetration with the second or saturation coat. Silica products typically do not have as high repellency rates as do silicone products, such as siloxanes, nor do they have as good weathering characteristics. (See Table 3.9.)

Sodium silicates

Sodium silicate materials should not be confused with water repellents. They are concrete densifiers or hardeners. Sodium silicates react with the free salts in concrete, such as calcium or free lime, making the concrete surface more dense. Usually these materials are sold as floor hardeners, which when compared to a true clear deck coating have repellency insufficient to be considered with materials of this section.

Water-Repellent Application

General surface preparations for all clear water-repellent applications require that the substrate be clean and dry. (Siloxane applications can be applied to slightly damp surfaces, but it is advisable to try a test application.) All release agents, oil, tar, and asphalt stains, as well as efflorescence, mildew, salt spray, and other surface contaminants, must be removed.

Application over wet substrates will cause either substrate discoloring, usually a white film formation, or water-repellent failure. When in doubt of moisture content in a substrate, a moisture test using a moisture meter or mat test using visquene taped to a wall to check for condensation is completed. Note that some silicone-based systems, such as silanes, must have moisture present, usually in the form of humidity, to complete the chemical reaction.

Substrate cracks are repaired before sealer application. Small cracks are filled with nonshrink grout or a sand–cement mixture. Large cracks or structural cracking should be epoxy injected. If a crack is expected to continue to move it should be sawn out to a minimum width of ¼ in and sealed with a compatible sealant.

Note that joint sealers should be installed first, as repellents contaminate joints causing sealant bonding failure. Concrete surfaces, including large crack patching, should be cured a minimum of 28 days before sealer application.

All adjacent substrates not being treated, including window frames, glass, and shrubberies, should be protected from overspray. Natural stone surfaces, such as limestone, are susceptible to staining by many clear sealers. Special formulations are available from manufacturers for these substrates. If any questions exist regarding an acceptable substrate for application, a test area should first be completed.

All sealers should be used directly from purchased containers. Sealers should never be thinned, diluted, or altered. Most sealers are recommended for application by low-pressure spray (20 lb/in^2), using a Hudson or garden-type sprayer. Brushes or rollers are also acceptable, but they reduce coverage rates. High-pressure spraying should only be used if approved by the manufacturer.

Applicators should be required to wear protective clothing and proper respirators, usually the cartridge type. Important cautionary measures should be followed in any occupied structure. Due to the solvents used in most clear sealers, application areas must be well ventilated. All intake ventilation areas must be protected or shut off to prevent the contamination of interior areas from sealer fumes. Otherwise, evacuation by building occupants is necessary.

Most manufacturers require a flood coating of material with coverage rates dependent upon the substrate porosity. Materials should be applied from the bottom of a building, working upward. Sealers are applied to produce a rundown or saturation of about 6 in of material below the application point for sufficient application. If a second coat is required, it should be applied in the same manner. Coverage rates for second coats increase as fewer materials will be required to saturate a substrate surface.

Testing should be completed to ensure that saturation of surfaces will not cause darkening or add sheen to substrate finishes. Dense concrete finishes may absorb insufficient repellent if they contain admixtures such as integral waterproofing or form release agents. In these situations, acid etching or pressure cleaning is necessary to allow sufficient sealer absorption. Approximate coverage rates of sealers over various substrates are summarized in Table 3.10.

Priming is not required with any type of clear sealer. However, some manufacturers recommend that two instead of one saturation coat be applied. Some systems may require a mist coat to break surface tension before application of the saturation coat.

TABLE 3.10 Coverage Rates for Water Repellents*

Surface	Coverage (ft/gal)
Steel-troweled concrete	150–300
Precast concrete	100–250
Textured concrete	100–200
Exposed aggregate concrete	100–200
Brick, dense	100–300
Brick, coarse	75–200
Concrete block, dense	75–150
Concrete block, lightweight	50–100
Natural stones	100–300
Stucco, smooth	125–200
Stucco, coarse	100–150

*Manufacturer's suggested rates should be referred to for specific installations. If a second coat is required, coverage will be higher for second application.

Cementitious Coatings

Cementitious-based coatings are among the oldest products used for above-grade waterproofing applications. Their successful use continues today even with numerous clear and elastomeric sealers available. However, cementitious systems have several disadvantages, including an inability to bridge cracks that develop in substrates after application. This can be nullified by installation of control or expansion joints to allow for movement. In remedial applications where all settlement cracks and shrinkage cracks have already developed, only expansion joints for thermal movement need be addressed.

These coatings are cement-based products containing finely graded siliceous aggregates that are nonmetallic. Pigments are added for color; proprietary chemicals are added for integral waterproofing or water repellency. An integral bonding agent is added to the dry mix, or a separate bonding agent liquid is provided to add to the dry packaged material during mixing. The cementitious composition allows use in both above- and below-grade applications. See Fig. 3.1 for a typical above-grade cementitious application.

Since these products are water resistant, they are highly resistant to freeze–thaw cycles; they eliminate water penetration that might freeze and cause spalling. Cementitious coatings have excellent color retention and become part of the substrate. They are also nonchalking in nature.

Figure 3.1 Above-grade cementitious application. (*Courtesy of Western Group*)

Color selections, such as white, that require the used of white Portland cement, increase material cost. Being cementitious, the product requires job site mixing, which should be carefully monitored to ensure proper in-place performance characteristics of coatings. Also, different mixing quotients will affect the dried finish coloring, and if each batch is not mixed uniformly, different finish colors will occur.

Cementitious properties

Cementitious coatings have excellent compressive strength, ranging from 4000 to 6000 lb/in^2 after curing (when tested according to ASTM C-109). Water absorption rates of cementitious materials are usually slightly higher than elastomeric coatings. Rates are acceptable for waterproofing and range from 3 to 5 percent maximum water absorption by weight (ASTM C-67).

Cementitious coatings are highly resistant to accelerated weathering, as well as being salt resistant. However, acid rain (sulfate contamination) will deteriorate cementitious coatings as it does other masonry products.

Cementitious coatings are breathable, allowing transmission of negative water vapor. This avoids the need for completing drying of substrates before application and the spalling that is caused by entrapped moisture. These products are suitable for the exterior of planters, undersides of balconies, and walkways, where negative vapor transmission is likely to occur. Cementitious coatings are also widely used on bridges and roads to protect exposed concrete from road salts, which can damage reinforcing steel by chloride attack.

Cementitious installations

Water entering masonry substrates causes brick to swell, which applies pressure to adjacent mortar joints. The cycle of swelling when wetted and relaxing when dry causes mortar joint deterioration. Cementitious coating application prevents water infiltration and the resulting deterioration. However, coatings alter the original facade aesthetics and a building owner or architect may deem them not acceptable.

Cementitious coatings are only used on masonry or concrete substrates, unlike elastomeric coatings that are also used on wood and metal substrates. Cementitious coating use includes applications to poured-in-place concrete, precast concrete, concrete block units, brick, stucco, and cement plaster substrates. Once applied, cementitious coatings bond so well to a substrate that they are considered an integral part of the substrate rather than a film protection such as an elastomeric coating.

Typical applications besides above-grade walls include swimming pools, tunnels, and retention ponds. With the Environmental Protection Agency (EPA) approval, these products may be used in water reservoirs and water treatment plants. Cementitious coatings are often used for finishing concrete, while at the same time providing a uniform substrate coloring.

An advantage with brick or block wall applications is that these substrates do not necessarily have to be tuck-pointed before cementitious coating application. Cementitious coatings will fill the voids, fissures, and honeycombs of concrete and masonry surfaces, effectively waterproofing a substrate.

When conditions require, complete coverage of the substrate by a process called bag, or face, grouting of the masonry is used as an alternative. In this process, a cementitious coating is brush applied to the entire masonry wall. At an appropriate time, the cementitious coating is removed with brushes or burlap bags, again revealing the brick and mortar joints. The only coating material left is that in the voids and fissures of masonry units and mortar joints. Although costly, this is an extremely effective means of waterproofing a substrate, more effective only than tuck pointing.

Complete cementitious applications provide a highly impermeable surface and are used to repair masonry walls that have been sandblasted to remove existing coatings and walls that are severely deteriorated. Cementitious applications effectively preserve a facade while making it watertight. Bag grouting application adds only a uniformity to substrate color; colored cementitious products can impart a different color to existing walls if desired. Mask grouting is similar to bag grouting. With mask grouting applications, existing masonry units are carefully taped over, exposing only mortar joints. The coating material is brush applied to exposed joints then cured. Tape is then removed from the masonry units, leaving behind a repaired joint surface with no change in wall facade color.

The thickness of coating added to mortar joints is variable but is greater when joints are recessed. This system is applicable only to substrates in which the masonry units themselves, such as brick, are nondeteriorated and watertight, requiring no restoration.

Texture is easily added to a cementitious coating, either by coarseness of aggregate added to the original mix or by application methods. The same cementitious mix applied by roller, brush, spray, hopper gun, sponge, or trowel results in many different texture finishes. This provides an owner or designer with many texture selections while maintaining adequate waterproofing characteristics. A summary of the

TABLE 3.11 Cementitious Coating Properties

Advantages	Disadvantages
Excellent bonding capability	No movement capability
Applicable to both above- and below-grade installations	Difficult to control uniform color and texture
Excellent weathering capabilities	High degree of expertise required for installation
Numerous textures and colors available	Not resistant to acid rain and other contaminants
Can eliminate need for tuck-pointing	Not applicable over wood or metal substrates

major advantages and disadvantages of cementitious coatings are given in Table 3.11.

In certain instances, such as floor–wall junctions, it is desirable first to apply the cementitious coating to a substrate, and then to fill the joint with sealant material in a color that matches the cementitious coating. The coating will fully adhere to the substrate and is compatible with sealant materials. It is also possible first to apply cementitious coating to substrates, then to apply a sealant to expansion joints, door, and window penetrations, and other joints. This is not possible with clear sealers nor recommended with elastomeric coatings due to bonding problems.

Cementitious coatings are a better choice over certain substrates, particularly concrete or masonry, than clear sealers or elastomeric coatings. This is because cementitious coatings have better bonding strength, a longer life cycle, lower maintenance, and less attraction of airborne contaminants. Providing adequate means are incorporated for thermal and structural movement, cementitious coatings will function satisfactorily for above- and below-grade waterproofing applications.

Cementitious Coating Application

For adequate bonding to substrates, surfaces to receive cementitious coatings should be cleaned of contaminants including dirt, efflorescence, form release agents, laitance, residues of previous coatings, and salts. Previously painted surfaces must be sandblasted or chemically cleaned to remove all paint film.

Cementitious coating bonding is critical to successful in-place performance. Therefore, extreme care should be taken in preparing substrates for coating application. Sample applications for bond strength should be completed if there is any question regarding the acceptability of a substrate, especially with remedial waterproofing applications.

Poured-in-place or precast concrete surfaces should be free of all honeycombs, voids, and fins. All tie holes should be filled before coating application with nonshrink grout material as recommended by the coating manufacturer. Although concrete does not need to be cured before cementitious coating application, it should be set beyond the green stage of curing. This timing occurs within 24 hours after initial concrete placement.

With smooth concrete finishes, such as precast, surfaces may need to be primed with a bonding agent. In some instances a mild acid etching can be desirable, using a muriatic acid solution and properly rinsing substrates before the coating application. Some manufacturers require a further roughing of smooth finishes, such as sandblasting, for adequate bonding.

On masonry surfaces, voids in mortar joints should be filled before coating installation. With both masonry and concrete substrates, existing cracks should be filled with a dry mix of cementitious material sponged into cracks. Larger cracks should be sawn out, usually to a ¾ in minimum, and packed with nonshrink material as recommended by the coating manufacturer.

Moving joints must be detailed using sealants designed to perform under the expected movement. These joints include thermal movement and differential movement joints. The cementitious material should not be applied over these joints as it will crack and "alligator" when movement occurs.

If cracks are experiencing active water infiltration, this pressure must be relieved before coating is applied. Relief holes should be drilled in a substrate, preferably at the base of the wall, to allow wicking of water, thus relieving pressure in the remainder of work areas during coating application. After application and proper curing time (approximately 48–72 hours), drainage holes may then be packed with a nonshrink hydraulic cement material and finished with the cementitious coating.

After substrate preparations are completed and just before application, substrates must be wetted or dampened with clean water for adequate bonding of the coating. Substrates must be kept continually damp in preparation for application. The amounts of water used are dependent on weather and substrate conditions. For example, in hot, dry weather, substrates require frequent wettings. Coatings should not be applied in temperatures below 40°F or in conditions when the temperature is expected to fall below freezing within 24 hours after application.

Cementitious coatings should be carefully mixed following the manufacturer's recommended guidelines concerning water ratios. Bonding agents should be added as required with no other additives or extenders, such as sand, used unless specifically approved by the manufacturer. With smooth surfaces such as precast concrete, an additional bonding agent is required.

Cementitious coatings may be applied by brush, trowel, or spray. Stiff, coarse, or fiber brushes are used for application. Brush applications require that the material be scrubbed into a substrate, filling all pores and voids. Finish is completed by brushing in one direction for uniformity.

Spray applications are possible by using equipment designed to move the material once mixed. Competent mechanics trained in the use of spray equipment and technique help ensure acceptable finishes and watertightness.

Trowel applications are acceptable for the second coat of material. Due to the application thickness of this method, manufacturers recommend that silica sand be added to the mix in proper portions. The first coats of trowel applications are actually brush applications that fill voids and pores. Finish trowel coats can be on a continuum from smooth to textured. Sponge finishing of the first coat is used to finish smooth concrete finishes requiring a cementitious application.

With textured masonry units such as split face or fluted block, additional material is required for effective waterproofing. On this type finish, spraying or brush applications are the only feasible and effective means.

The amounts of material required depends upon the expected water conditions. Under normal waterproofing requirements, the first coat is applied at a rate of 2 lb of material per square yard of work area. The finish coat is then applied at a coverage rate of 1 lb/yd^2. In severe water conditions, such as below-grade usage with water-head pressures, materials are applied at 2 lb/yd^2. This is followed by a trowel application at 2 lb/yd^2. Clean silica sand is added to the second application at 25 lb of silica to one bag, 50 lb, of premixed cementitious coating.

With all applications, the second material coat should be applied within 24 hours after applying the first coat. Using these application rates, under normal conditions, a 50-lb bag of coating will cover approximately 150 ft^2 (1 lb/yd^2, first coat; 2 lb/yd^2, second coat). The finish thickness of this application is approximately ⅛ in.

Trying to achieve this thickness in one application or adding excessive material thickness in one application should not be attempted. Improper bonding will result, and material can become loose and spall. To eliminate mortar joint shadowing on a masonry wall being visible through the coating, a light trowel coat application should be applied first, followed by a regular trowel application.

The cementitious coating beginning to roll or pull off a substrate is usually indicative of the substrate being too dry; redampening with clean water before proceeding is necessary. Mix proportions must be kept constant and uniform, or uneven coloring or shadowing of the substrate will occur.

After cementitious coatings are applied they should be cured according to the manufacturer's recommendations. Typically, this requires keeping areas damp for 1–3 days. In extremely hot weather, more frequent and longer cure times are necessary to prevent cracking of the coating. The water cure should not be done too soon after application as it may ruin or harm the coating finish. Chemical curing agents should not be used or added to the mix unless specifically approved by the coating manufacturer.

Typically, primers are not required for cementitious coating applications, but bonding agents are usually added during mixing. In some cases, if substrates are especially smooth or previous coatings have been removed, a direct application of the bonding agent to substrate surfaces is used as a primer. If there is any question regarding bonding strength, samples should first be applied both with and without a bonding agent and tested before proceeding with the complete application.

Cementitious coatings should not be applied in areas where thermal, structural, or differential movement will occur. Coatings will crack and fail if applied over sealant in control or expansion joints. Cementitious-based products should not be applied over substrates other than masonry substrates such as wood, metal, or plastics.

Elastomeric Coatings

Paints and elastomeric coatings are similar in that they always contain three basic elements in a liquid state—pigment, binder, and solvent. In addition, both often contain special additives such as mildew-resistant chemicals. However, paints and coatings differ in their intended uses.

Paints are applied only to add decorative color to a substrate. Coatings are applied to waterproof or otherwise protect a substrate. The difference between clear sealers and paints or coatings is that sealers do not contain the pigments that provide the color of paints or coatings.

Solvent is added to paints and coatings to lower the material viscosity so it can be applied to a substrate by brush, spray, or roller. The binder and solvent portion of a paint or coating is referred to as the vehicle. A coating referred to as 100 percent solids is merely a binder in a liquid state that cures, usually moisture cured from air humidity, to a seamless film upon application. Thus, it is the binder portion, common to all paints and coatings, that imparts the unique characteristics of the material, differentiating coatings from paints.

Waterproof coatings are classified generically by their binder type. The type of resin materials added to the coating imparts the waterproofing characteristics of the coating material. Binders are present in the vehicle portion of a coating in either of two types. An emulsion occurs when binders are dispersed or suspended in solvent for purposes of application. Solvent-based materials have the binder dissolved within the solvent.

The manner in which solvents leave a binder after application depends upon the type of chemical polymer used in manufacturing. A thermoplastic polymer coating dries by the solvent evaporating and leaving behind the binder film. This is typical of water-based acrylic elastomeric coatings used for waterproofing. A thermosetting polymer reacts chemically or cures with the binder and can become part of the binder film that is formed by this reaction. Examples are epoxy paints, which require the addition and mixing of a catalyst to promote chemical reactions for curing the solvent.

The catalyst prompts a chemical reaction that limits application time for these materials before they cure in the material container. This action is referred to as the "pot life" of material (workability time). The chemical reactions necessary for curing create thermosetting polymer vehicles that are more chemically resistant than thermoplastic materials. Thermosetting vehicles produce a harder film and have an ability to contain higher solids content than thermoplastic materials.

Resins used in elastomeric coatings are breathable. They allow moisture vapor transmission from the substrate to escape through the coating without causing blisters in the coating film. This is a favorable characteristic for construction details at undersides of balconies that are subjected to negative moisture drive. Thermosetting materials such as epoxy paints are not breathable. They will blister or become unbonded from a substrate if subjected to negative moisture drive.

Resins

Elastomeric coatings are manufactured from acrylic resins with approximately 50 percent solids by volume. Most contain titanium dioxide to prevent chalking during weathering. Additional additives include mildewcides, alkali-resistant chemicals, various volume extenders to increase solids content, and sand or other fillers for texture.

Resins used in waterproofing coatings must allow the film to envelop a surface with sufficient dry film millage (thickness of paint measured in millimeters) to produce a film that is watertight, elastic, and breathable. Whereas paints are typically applied 1–4 mil thick, elastomeric coatings are applied 10–20 mil thick.

It is this thickness (with the addition of resins or plasticizers that add flexibility to the coating) that creates the waterproof and elastic coating, thus the term *elastomeric coating*. Elastomeric coatings have the ability to elongate a minimum of 300 percent at dry millage thickness of 12–15 mil. Elongation is tested as the minimum ability of a coating to expand then return to its original shape with no cracking or splitting (tested according to ASTM D-2370). Elongation should be tested after aging and weathering to check effectiveness after exposure to the elements.

Elastomeric coatings are available in both solvent-based and water-based vehicles. Water-based vehicles are simpler to apply and not as moisture sensitive as solvent based. The latter is applied only to totally dry surfaces that require solvent materials for cleanup.

Typical properties of a high-quality, waterproof, and elastic coating include

- Minimum of 10-mil dry application
- High solids content (resins)
- Good ultraviolet weathering resistance
- Low water absorption, withstanding hydrostatic pressure
- Permeability for vapor transmission
- Crack bridging capabilities
- Resistance to sulfites (acid rain) and salts
- Good color retention and low dirt pickup
- High alkali resistance

Acrylic coatings are extremely sensitive to moisture during their curing process, taking up to 7 days to cure. Should the coating be subjected to moisture during this time, it may reemulsify (return to liquid state). This becomes a critical installation consideration whenever such coatings are used in a horizontal or slightly inclined surface that might be susceptible to ponding water.

Elastomeric coating installations

Elastomeric coatings, which are used extensively on stucco finish substrates and exterior insulation finish systems (EIFS), are also used on masonry block, brick, concrete, and wood substrates. Some are available with asphalt primers for application over asphalt finishes. Others have formulations for use on metal and sprayed urethane foam roofs.

TABLE 3.12 Elastomeric Coating Properties

Advantages	Disadvantages
Excellent elastomeric and crack bridging capability	Uniform application thickness difficult to control
Wide range of colors and textures available	Life cycle shorter than cementitious
Breathable	No below-grade usage
Applicable over wood and metal substrates	Masonry substrates may require extensive repairs before application
Resistant to acid rain and other pollutants	May fade over time

Elastomeric coatings are also successfully used over previously painted surfaces. By cleaning, preparing the existing surface, repairing cracks, and priming, coatings can be used to protect concrete and masonry surfaces that have deteriorated through weathering and aging.

Proper preparation, such as tuck-pointing loose and defective mortar joints and injecting epoxy into cracks, must be completed first. In single wythe masonry construction, such as split face block, applying a cementitious block filler is necessary to fill voids in the block before applying elastomeric coating for effective waterproofing.

Aesthetically, coatings are available in a wide range of textures and are tintable to any imaginable color. However, deep, dark, tinted colors may fade, or pigments added for coloring may bleed out creating unsightly staining. Heavy textures limit the ability of a coating to perform as an elastomeric due to the amount of filler added to impart texture. Because elastomeric coatings are relatively soft materials (lower tensile strength to impart flexibility), they tend to pick up airborne contaminants. Thus lighter colors, including white, may get dirty quickly.

Uniform coating thickness is critical to ensure crack bridging and thermal movement capabilities after application. Applicators should have wet millage gages for controlling the millage of coating applied. Applications of elastomeric coatings are extremely labor sensitive. They require skilled application of the material. In addition, applicators must transition coating applications into adjacent members of the building envelope, such as window frames and flashings, for effective envelope waterproofing. (See Table 3.12.)

Elastomeric Coating Application

Successful application of elastomeric coatings depends entirely on proper substrate preparation. Although they are effective waterproof materials, they should not be applied over cracks, voids, or deteriorated materials, as this will prevent cohesive waterproofing of the building envelope. Coatings chosen must be compatible with any existing coatings, sealants, or patching compounds used in crack repairs. Coating manufacturers have patching, sealing, and primer materials, all compatible with their elastomeric coating.

Applying elastomeric coating requires applicator knowledge beyond a typical paint job. Most painting contractors do not have the experience or knowledge to apply these coatings.

Existing substrates must be cleaned to remove all dirt, mildew, and other contaminants. This is accomplished by pressure cleaning equipment with a minimum capability of 1500 lb/in^2 water pressure. All grease, oils, and asphaltic materials must be removed completely.

Mildew removal with chlorine should be done where necessary. Chemical cleaning is also necessary to remove traces of release agents or incompatible curing agents. If chemicals are used, the entire surface should be rinsed to remove any chemical traces that might affect the coating bonding.

Previously painted substrates should have a duct-tape test for compatibility of the elastomeric coating application. A sample area of coating should be applied over existing materials and allowed to dry. Then duct tape should be sealed firmly to the substrate then pulled off quickly. If any amount of coating comes off with the tape, coatings are not properly adhering to existing materials. In that case, all existing coatings or paints must be removed to ensure adequate bonding. No coating can perform better than the substrate to which it is applied, in this case a poorly adhered existing coating. Excessively chalky coatings either must be removed or a primer coat applied. Primers will effectively seal the surface for proper bonding to a substrate.

High alkaline masonry substrates must be checked for a pH rating before installation. The pH rating is a measure of substrate acidity or alkalinity. A rating of 7 is neutral, with higher ratings corresponding to higher alkaline substrates. A pH of more than 10 requires following specific manufacturer's recommendations. These guidelines are based upon the alkali resistance of a coating and substrate pH.

Surface preparations of high alkali substrates include acid washing with 5 percent muriatic acid or primer application. In some cases, extending curing time of concrete or stucco substrates will effectively lower their pH. Immediately after application stucco has a high pH, but it has continually lower pH values during final curing stages. New stucco should cure for a minimum of 30 days, preferably 60–90 days, to lower the pH. This also allows shrinkage and thermal cracks to form and be treated before coating application.

Sealant installation should be completed before applying elastomeric coating to prevent joint containment by the coating. This includes expansion and control joints, perimeters of doors and windows, and flashings. Small nonmoving cracks less than 1/16 in wide require filling and overbanding 2 in wide with a brushable or knife-grade sealant material.

Cracks exceeding 1/16 in that are also nonmoving joints should be sawn out to approximately 1/4 in in width and depth and filled with a knife-grade sealant, followed by overbanding approximately 4 in wide (see Fig. 3.2).

Overbanding (bandage application of a sealant) requires skilled craftspeople to featheredge banding sides to prevent telescoping of patches through the coating. Thick, unfeathered applications of brushable sealant will show through coating applications, providing an unacceptable substrate appearance.

Large cracks over 1/2 in wide that are nonmoving, such as settlement cracks, should be sawn out and proper backing materials applied before sealant installation. Fiberglass mesh in 4 in widths can be embedded into the brushable sealant for additional protection.

Joints that are expected to continue moving, such as joints between dissimilar materials, should be sealed using guidelines set forth in Chap. 4. These joints should not be coated over, since the movement

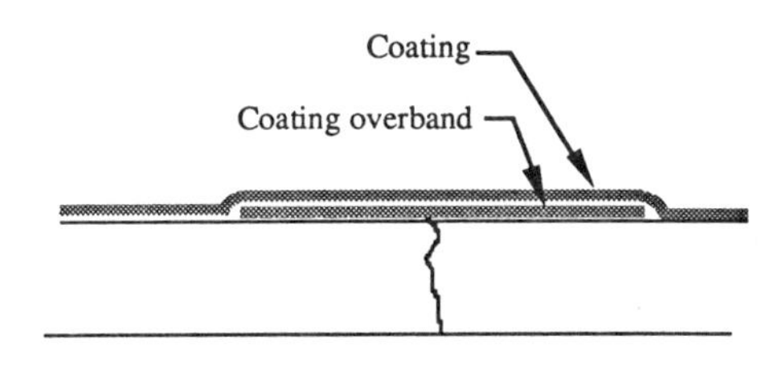

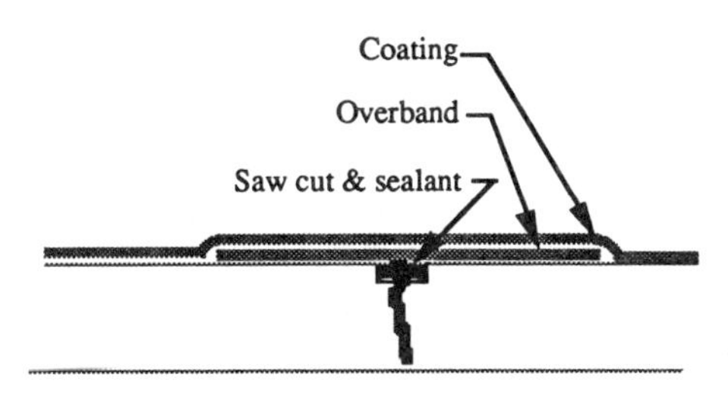

Figure 3-2 Elastomeric crack-repair detailing.

experienced at these joints typically exceeds the elastomeric coating capability. In such cases, the coating will alligator and develop an unsightly appearance.

Brick or block masonry surfaces should be checked for loose and unbonded mortar joints. Faulty joints should be tuck-pointed or sealed with a proper sealant. With masonry applications, when all mortar joints are unsound or excessively deteriorated, all joints should be sealed before coating.

Additionally, with split face block, particularly single wythe construction, a cementitious block filler should be applied to all cavities and voids. This provides the additional waterproofing protection that is necessary with such porous substrates. On previously painted split face construction, an acrylic block filler may be used to prepare the surface.

All sealants and patching compounds must be cured before coating application; if this is not done, patching materials will mildew beneath the coating and cause staining. For metal surfaces, rusted portions must be removed or treated with a rust inhibitor then primed as recommended by the coating manufacturer. New galvanized metal should also be primed.

Wood surfaces require attention to fasteners that should be recessed and sealed. Laps and joints must also be sealed. Wood primers are generally required before coating application. The success of an elastomeric coating can depend upon use of a proper primer for specific conditions encountered. Therefore, it is important to refer to manufacturer guidelines for primer usage.

Elastomeric coatings are applied by brush, roller, or spray after proper mixing and agitating of the coating (see Fig. 3.3). Roller application is preferred as it fills voids and crevices in a substrate. Long nap rollers should be used with covers having a ¾–1½-in nap. Elastomeric coatings typically require two coats to achieve proper millage. The first application must be completely dried before the second coat is applied.

Spray applications require a mechanic properly trained in the crosshatch method. This method applies coating by spraying vertically and then horizontally to ensure uniform coverage. Coatings are then back rolled with a saturated nap roller to fill voids and crevices.

Brushing is used to detail around windows or protrusions, but it is not the preferred method for major wall areas. When using textured elastomeric coatings, careful application is extremely important to prevent unsightly buildup of texture by rolling over an area twice. Placing too much pressure on a roller nap reduces the texture applied and presents an unsightly finish. Textured application should not be rolled over adjacent applications as roller seams will be evident after drying.

Figure 3.3 Elastomeric coating application after preparatory work is completed. (*Courtesy of Innovative Coatings*)

Coatings, especially water-based ones, should not be applied in temperatures lower than 40°F and should be protected from freezing by proper storage. Manufacturers do not recommend application in humidity over 90 percent. Application over excessively wet substrates may cause bonding problems. In extremely hot and dry temperatures, substrates are misted to prevent premature coating drying. Complete curing takes 24–72 hours; coatings are usually dry to the touch and ready for a second coat in 3–5 hours, depending on the weather.

Coverage rates vary depending upon the substrate type, porosity of the substrate, and millage required. Typically, elastomeric coatings are applied at 100–150 ft^2/gal per coat, for a net application of 50–75 ft^2/gal. This results in a dry film thickness of 10–12 mil.

Elastomeric coatings should not be used in below-grade applications where they can reemulsify and deteriorate, nor are they designed for horizontal surfaces subject to traffic. Horizontal areas such as copings or concrete overhangs should be checked for ponding water that may cause debonding and coating reemulsification.

Deck Coatings

Several choices are available for effective waterproofing of horizontal portions of a building envelope. Several additional choices of finishes or wearing surfaces over this waterproofing are also available. Liquid-applied seamless deck coatings or membranes are used where normal roofing materials are not practical or acceptable. Deck coatings may be applied to parking garage floors, plaza decks, balcony decks, stadium bleachers, recreation roof decks, pool decks, observation decks, and helicopter pads. In these situations, coatings waterproof occupied areas beneath the decks and provide wearing surfaces acceptable for either vehicular or pedestrian traffic. These systems do not require topping slabs or protection such as tile pavers to protect them from traffic.

Deck coatings make excellent choices for remedial situations where it is not possible to allow for the addition of a topping slab or other waterproofing system protection. Deck coatings are installed over concrete, plywood, or metal substrates but should not be installed over lightweight insulating concrete.

Deck coatings are also used to protect concrete surfaces from acid rain, freeze–thaw cycles, and chloride ion penetration, and to protect reinforcing steel.

In certain situations, deck coatings are not specifically installed for their waterproofing characteristics but for protection of concrete against environmental elements. For example, whereas deck coatings on the first floor of a parking garage protect occupied offices on ground level, they also protect concrete against road salts and freeze–thaw cycles on all other levels. In these situations, coatings are installed to prevent unnecessary maintenance costs and structural damage during structure life cycling.

Deck coatings are usually installed in two- to four-step applications with the final coat containing aggregate or grit to provide a nonslip wearing surface for vehicular or foot traffic. Aggregate is usually broadcast into the final coat either by hand seeding or by mechanical spray such as sandblast equipment. Aggregates include silica sand, quartz carbide, aluminum oxide, or crushed walnut shells. The softer, less harsh silica sand is used for pedestrian areas; the harder wearing aggregate is used for vehicular traffic areas. The amount of aggregate used varies, with more grit concentrated in areas of heavy traffic such as parking garage entrances or turn lanes.

Due to the manufacturing processes involved, deck coatings are available in several standard colors but usually not in custom colors. A standard gray color is recommended for vehicular areas because oils and tire trackings will stain lighter colors. Some manufacturers allow their coatings to be color top coated with high-quality urethane coatings, if a special color is necessary, but only in selected cases and not in vehicular areas.

Deck coatings are supplied in two or three different formulations for base, intermediate, and wearing coats. Base coats are the most elastomeric of all formulations. Since they are not subject to wear, they do not require the high tensile strength or impact resistance that wearing layers require. Lower tensile strength allows a coating to be softer and, therefore, to have more elastomeric and crack bridging characteristics than top coats. As such, base coats are the waterproof layer of deck-coating systems.

Top and intermediate coats are higher in tensile strength and are impact resistant to withstand foot or vehicular traffic. However, the various coating layers must be compatible and sufficiently similar to base coat properties not to crack or alligator as a paint applied over an elastomeric coating might. This allows base coatings to move sufficiently to bridge cracks that develop in substrates without cracking top coats.

Adding grit or aggregate in a coating further limits movement capability of top coats. The more aggregate added, the less movement top coats can withstand, further restricting movement of base coats.

Deck coatings are available in several different chemical formulations. They are differentiated from clear coatings, which are penetrating sealers, in that they are film-forming surface sealers. Deck coating formulations include

- Acrylics
- Cementitious coatings
- Epoxy
- Asphalt overlay
- Latex
- Neoprene
- Hypalon
- Urethane
- Modified urethane

Acrylics

Acrylics are not waterproof coatings but act as water-repellent sealers. Their use is primarily aesthetic, to cover surface defects and cracking in decks. These coatings have low elastomeric capabilities; silica aggregate is premixed directly into their formulations, which further lowers their elastic properties. These two characteristics prevent acrylics from being true waterproof coatings.

The inherent properties of acrylics protect areas such as walkways or balconies with no occupied areas beneath from water and chloride penetration. In addition to concrete substrates, acrylics are used over wood or metal substrates provided recommended primers are installed.

TABLE 3.13 Acrylic Deck-Coating Properties

Advantages	Disadvantages
Ease of application	Not a complete waterproof system
Aggregate is integral with coating	No movement capability
Slab-on-grade applications	Not resistant to vehicular traffic

Acrylics are also used at slab-on-grade areas where urethane coatings are not recommended.

Sand added in acrylic deck coatings provides excellent antislip finishes. As such, they are used around pools or areas subject to wet conditions that require protection against slips and falls. Acrylics are not recommended for areas subject to vehicular traffic. Some manufacturers allow their use over asphaltic pavement subject only to foot traffic for aesthetics and a skid-resistant finish. (See Table 3.13.)

Cementitious

Cementitious deck coatings are used for applications over concrete substrates and include an abrasive aggregate for exposure to traffic. These materials are supplied in prepacked and premixed formulations requiring only water for mixing. Cementitious coatings are applied by trowel, spray, or squeegee, the latter being a self-leveling method.

Cementitious systems contain proprietary chemicals to provide necessary bonding and waterproofing characteristics. These are applied to a thickness of approximately ⅛ in and will fill minor voids in a substrate. A disadvantage of cementitious coatings, like below-grade cementitious systems, is their inability to withstand substrate movement or cracking. They are one-step applications, with an integral wearing surface, that require no primers and are applicable over damp concrete surfaces.

Modified acrylic cementitious coatings are also available. Such systems typically include a reinforcing mesh embedded into the first coat to improve crack-bridging capabilities. Acrylics are added to the basic cement and sand mixture to improve bonding and performance characteristics.

Cementitious membrane applications include the dry-shake and power-trowel methods previously discussed in Chap. 2. Successful applications depend on properly designed, detailed, and installed allowances for movement, both thermal and differential. For cementitious membranes to be integrated into a building envelope, installations should include manufacturer-supplied products for cants, patching, penetrations, and terminations. (See Table 3.14.)

TABLE 3.14 Cementitious Deck-Coating Properties

Advantages	Disadvantages
Excellent bonding to concrete substrates	No movement capabilities
Good wearing surfacing	Not applicable over wood or metal
Dry-shake and power-trowel applications	Not resistant to acid rain and other contamination

Epoxy

As with acrylics, epoxy coatings are generally not considered true waterproof coatings. They are not recommended for exterior installations due to their poor resistance to ultraviolet weathering. Epoxy floor coatings have very high tensile strengths, resulting in low elastomeric capabilities. These coatings are very brittle and will crack under any movement, including thermal and structural.

Epoxy coatings are used primarily for interior applications subject to chemicals or harsh conditions such as waste and water treatment plants, hospitals, and manufacturing facilities. For interior applications not subject to movement, epoxy floor coatings provide effective waterproofing at mechanical room floor, shower, and locker room applications. Epoxy coatings are available in a wide variety of finishes, colors, and textures and may be roller or trowel applied.

Epoxy deck coatings are also used as top coats over a base-coat waterproof membrane of urethane or latex. However, low-movement capabilities and brittleness of epoxy coatings limit elastomeric qualities of waterproof top coats. (See Table 3.15.)

Asphalt

Asphalt overlay systems provide an asphalt wearing surface over a liquid-applied membrane. The waterproofing base coat is a rubberized asphalt or latex membrane that can withstand the heat created during installation of the asphalt protective course. Both the waterproof membrane and the asphalt layers are hot-applied systems.

Asphalt layers are approximately 2 in thick. These systems have bet-

TABLE 3.15 Epoxy Deck-Coating Properties

Advantages	Disadvantages
Excellent chemical resistance	Brittle; no movement capability
High tensile strength	Trowel application
Variety of finishes, colors, and textures	Not for exterior applications

TABLE 3.16 Asphalt Deck-Coating Properties

Advantages	Disadvantages
Protection of membrane by asphalt overlay	Weight added to structure
Longer wearing capability	Movement capability restricted
Thickness of applied system	Inaccessibility for repairs

ter wearing capabilities due to the asphaltic overlay protecting the waterproof base coating.

The additional weight added to a structure by these systems must be calculated to ensure that an existing parking garage can withstand the additional dead loads that are created. Asphalt severely restricts the capability of the waterproof membrane coating to bridge cracks or to adjust to thermal movement. Additionally, it is difficult to repair the waterproofing membrane layers once the asphalt is installed. There is no way to remove overlays without destroying the base coat membrane. Asphaltic systems are not recoatable. For maintenance, they must be completely removed and reinstalled. (See Table 3.16.)

Latex, neoprene, hypalon

Deck coatings are available in synthetic rubber formulations, including latex, neoprene, neoprene cement, and hypalon. These formulations include proprietary extenders, pigments, and stabilizers. Neoprene derivatives are soft, low-tensile materials and require the addition of a fabric or fiberglass reinforcing mesh. For traffic-wear resistance, this reinforcing mesh enhances in-place performance properties such as elongation and crack-bridging capabilities. Reinforcing requires that the products be trowel applied rather than roller or squeegee applied.

Trowel application and a finish product thickness of approximately ¼ in increase the in-place costs of these membranes. They also require experienced mechanics to install the rubber derivative systems. Trowel applications, various derivatives, and proprietary formulations provide designers with a wide range of textures, finishes, and colors.

Rubber compound coatings have better chemical resistance than most other deck-coating systems. They are manufactured for installation in harsh environmental conditions such as manufacturing plants, hospitals, and mechanical rooms. They are appropriate in both exterior and interior applications.

Design allowances must be provided for finished application thickness. Deck protrusions, joints, wall-to-floor details, and equipment supports must be flashed and reinforced for membrane continuity and

TABLE 3.17 Latex, Neoprene, and Hypalon Deck-Coating Properties

Advantages	Disadvantages
Excellent chemical resistance	Trowel application required
Good aging and weathering	Fabric reinforcement required
Good wear resistance	High cost

watertightness. Certain derivatives of synthetic rubbers become brittle under aging and ultraviolet weathering, which hinders waterproofing capabilities after installation. Manufacturer's literature and applicable test results should be reviewed for appropriate coating selection. (See Table 3.17.)

Urethanes

Urethane deck coatings are frequently used for exterior deck waterproofing. These are available for both pedestrian and vehicular areas in a variety of colors and finishes. Urethane systems include aromatic, aliphatic, and epoxy-modified derivatives and formulations.

Aliphatic materials have up to three times the tensile strength of aromatics but only 50 percent of aromatic elongation capability. Many manufacturers use combinations of these two materials for their deck-coating systems. Aromatic materials are installed as base coats for better movement and recovery capabilities; aliphatic urethane top coats make for better weathering, impact resistance, and ultraviolet resistance.

Epoxy urethane systems are also used as top coat materials. These modified urethane systems provide additional weathering and wear, while still maintaining necessary waterproofing capabilities.

Urethane coatings are applied in two or more coats, depending upon the expected traffic wear. Aggregate is added in the final coating for a nonslip wearing surface. An installation advantage with urethane systems is their self-flashing capability. Liquid-applied coatings by brush application are turned up adjoining areas at wall-to-floor junctions, piping penetrations, and equipment supports and into drains.

Urethane coatings are manufactured in self-leveling formulations for applications control of millage on horizontal surfaces. Nonflow or detailing grades are available for vertical or sloped areas. The uncured self-leveling coating is applied by notched squeegees to control thickness on horizontal areas. At sloped areas, such as the up and down ramps of parking garages or vertical risers of stairways, nonflow material application ensures proper millage. If self-leveling grade is used in

TABLE 3.18 Urethane Deck-Coating Properties

Advantages	Disadvantages
Excellent crack-bridging capability	Limited color selection
Simple installations	Low chemical resistance
Expanded product line	Maintenance required with heavy traffic

these situations, material will flow downward and insufficient millage at upper areas of the vertical or sloped portions will occur.

Nonflow liquid material is used to detail cracks in concrete decks before deck-coating application. Cracks wider than 1/16 in, which is the maximum width urethane materials bridge without failure, are sawn out and sealed with a urethane sealant. This area is then detailed 4 in wide with nonflow coating.

In addition, urethane coatings are compatible with urethane sealants used for cants between vertical and horizontal junctions, providing a smooth transition in these and other changes of plane. This is similar to using wood cants for roof perimeter details (see Table 3.18).

Deck-Coating Characteristics

Deck coatings bond directly to concrete, wood, or metal substrates. This prevents lateral movement of water beneath the coatings as is possible with sheet good systems. Once cured, coatings are nonbreathable and blister if negative vapor drive is present. This is the reason deck coatings, with the exception of acrylic and epoxies, are not recommended for slab-on-grade applications. Specifically, moisture in soils is drawn up into a deck by capillary action, causing blistering in applied deck coatings. In the same manner, blistering occurs in deck coatings applied on upper deck portions of sandwich-slab membranes due to entrapped moisture and negative vapor drive. In both cases, an epoxy vapor barrier prime coat should be installed to protect deck-coating systems from being subjected to this vapor drive.

Physical properties of deck coatings vary as widely as the number of systems available. Important considerations to review when choosing a coating system include tensile strength, elongation, chemical resistance, weathering resistance, and adhesion properties. Different installation types, expected wearing, and weathering conditions require different coating types.

High tensile strength is necessary when a coating is subject to heavy wear including vehicular traffic or forklift traffic at loading docks. Tensile strengths of some deck coatings exceed 1000 lb/in^2 (tested according to ASTM D-412) and are higher for epoxy coatings. This high tensile strength reduces the elongation ability of coatings.

Elongation properties range from 200 percent (for high tensile strength top coats) to more than 1000 percent (for low tensile strength base coats). For pedestrian areas where impact resistance and heavy wear is not expected, softer, higher elongation aromatic urethanes are used. Sun decks subject to impact from lawn chairs and tables would better be served by a coating between the extremes of high and low tensile strength.

Chemical resistance can be an important consideration under certain circumstances. Parking garage decks must have coatings resistant to road salts, oil, and gasoline. A pedestrian sun deck may be subjected to chlorine and other pool chemicals. Testing for chemical resistance should be completed according to recognized tests such as ASTM D-471.

Weathering resistance and ultraviolet resistance are important to coatings exposed to the elements such as on upper levels of a parking garage. These areas should be protected by the ultraviolet-resistant properties of coatings such as an aliphatic urethane. Weathering characteristics can be compared with accelerated weathering tests such as ASTM D-822. Other properties to consider on an as-needed basis include adhesion tests, solvent odor for interior uses, moisture vapor transmission, and fire resistance.

Once installed, the useful life of deck coatings depends upon proper maintenance as well as traffic wear. Heavily traveled parking garage decks and loading docks will wear faster than a seldom used pedestrian deck area. To compensate, manufacturers recommend a minimum of one to as many as three additional intermediate coat applications. Additional aggregate is also added for greater wear resistance.

With proper installation, deck coatings should function for upward of 5 years before requiring resealing. Resealing entails cleaning, patching existing coatings as required, reapplying top coatings, and, if required, adding intermediate coats at traffic lanes. Proper maintenance prevents coatings from being worn and exposing base coatings that cannot withstand traffic or exposure.

Exposed and unmaintained deck-coating systems require complete removal and replacement when repairs become necessary. Chemical spills, tears or ruptures, and improper usage must also be repaired to prevent unnecessary coating damage. Maintaining the top coat or wearing surface properly will extend the life cycle of a deck-coating system indefinitely.

Deck coatings are also effective in remedial waterproofing applications. If a sandwich-slab membrane installed during original construction becomes ineffective, a deck coating can be installed over the topping slab provided proper preparatory work is completed. Deck coatings can also be successfully installed over quarry and other hard finish tile surfaces, precast concrete pavers, and stonework. With any special surfacing installation, proper adhesive tests and sample applications should be completed.

Deck-Coating Application

Substrate adhesion and proper substrate finishing are critical for successful deck-coating applications. In general, substrates must be clean, dry, and free of contaminants. Concrete substrates exhibiting oil or grease contamination should be cleaned with a biodegradable degreaser such as trisodium phosphate. Contaminants such as parking stall stripe paint should be removed by mechanical grinder or sandblasting.

For new concrete substrates, a light broom finish is desirable. Surface laitance, fins, and ridges must be removed. Honeycomb and spalled areas should be patched using an acceptable nonshrink grout material.

Coatings should not be applied to exposed aggregate or reinforcing steel. If present, these areas should be properly repaired. Concrete surfaces, including patches, should be cured a minimum of 21 days before coating application. Use of most curing compounds is prohibited by coating manufacturers since resins contained in curing compounds prevent adequate adhesion. If present, substrates require preparatory work, including sandblasting, or acid etching with muriatic acid. Water curing is desirable, but certain manufacturers allow use of sodium silicate curing agents.

Substrate cracks must be prepared before coating application. Cracks less than 1/16 in wide should be filled and detailed with a 4-in band of nonflow base coat. Larger cracks, from ½ in to a maximum of 1 in width, should be sawn out and filled with urethane sealant. Moving joints should have proper expansion joints installed with coating installed up to but not over these expansion joints. Refer to Fig. 3.4 for typical crack detailing.

Substrates should be sloped to drain water toward scuppers or deck drains. Plywood surfaces should be swept clean of all dirt and sawdust. Plywood should be of A-grade only, with tongue and groove connections. Only screw-type fasteners should be used, and they should be countersunk. The screw head is filled with a urethane sealant and troweled flush with the plywood finish. As these coatings are relatively thin, 60–100 mil dry film, their finish mirrors the substrate they are applied over. Therefore, if plywood joints are uneven or knots or chips

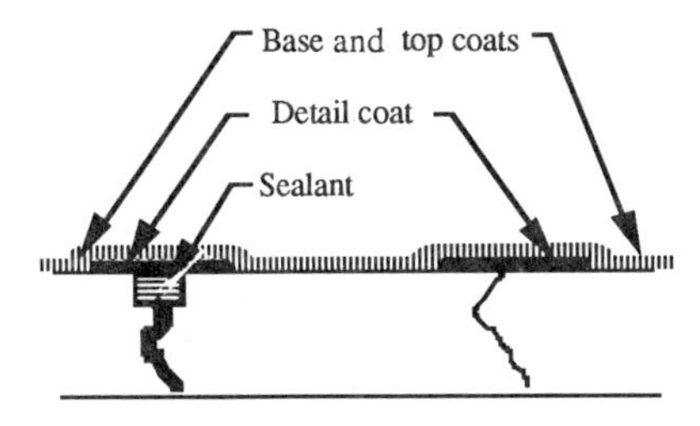

Figure 3.4 Crack detailing.

are apparent in the plywood, they also will be apparent in the deck-coating finish.

Metal surfaces require sandblasting or wire brush cleaning, then priming immediately afterward. Aluminum surfaces also require priming. Other substrates such as PVC, quarry tile, and brick pavers should be sanded to roughen the surface for adequate adhesion. Sample test areas should be completed to check adhesion on any of these substrates before entire application.

For recoating over previously applied deck coatings, existing coatings must be thoroughly cleaned with a degreaser to remove all dirt and oil. Delaminated areas should be cut out and patched with base coat material. Before reapplication of top coats, a solvent is applied to reemulsify existing coatings for bonding of new coatings.

All vertical abutments and penetrations should be treated by installing a sealant cover, followed by a detail coat of nonflow material. If a joint occurs between changes in plane such as wall-to-floor joints, a fabric reinforcing of neoprene or fiberglass should be embedded into the detail coat for reinforcement. Figure 3.5 shows typical installation procedures for this work.

With new construction, detail coats of base coat membrane are

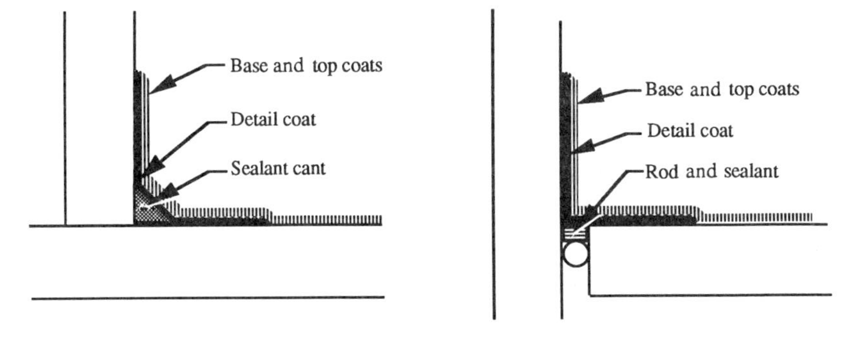

Figure 3.5 Floor-to-wall juncture detailing.

turned up behind the facing material (e.g., brick cavity wall), followed by coating and detailing to the facing material. This allows for double protection in these critical envelope details. At doors or sliding glass doors, coating is installed beneath thresholds before installation of doors.

For applications over topping slabs with precast plank construction, such as double-T, a joint should be scored at every T-joint. These joints are then filled with sealant and a detail coat of material is applied, allowing for differential and thermal movement. Refer to Fig. 3.6 for a typical installation detail at these areas.

Base coats are installed by notched squeegees for control of millage, typically 25–40 mil dry film, followed by back rolling of materials for uniform millage thickness. Following initial base-coat curing, within 24 hours intermediate coats, top coats, and aggregates are installed. (See Fig. 3.7.)

Aggregate, silica sand, and silicon carbide are installed in intermediate or final top coats or possibly both in heavy traffic areas. On pedestrian decks grit is added at a rate of 4–10 lbsquare (100 ft^2) of deck area. In traffic lanes, as much as 100–200 lb of aggregate per square is added.

Aggregate is applied by hand seeding (broadcasting) or by mechanical means (sandblast equipment). If aggregate is added to a top coat, it is back rolled for uniform thickness of membrane and grit distribution. With installations of large aggregate amounts, an initial coat with aggregate fully loaded is first allowed to dry. Excess aggregate is then swept off, and an additional top coat is installed to lock in the grit and act as an additional protective layer. See Fig. 3.8 for aggregate comparisons.

Intermediate coats usually range in thickness from 10–30 mil dry

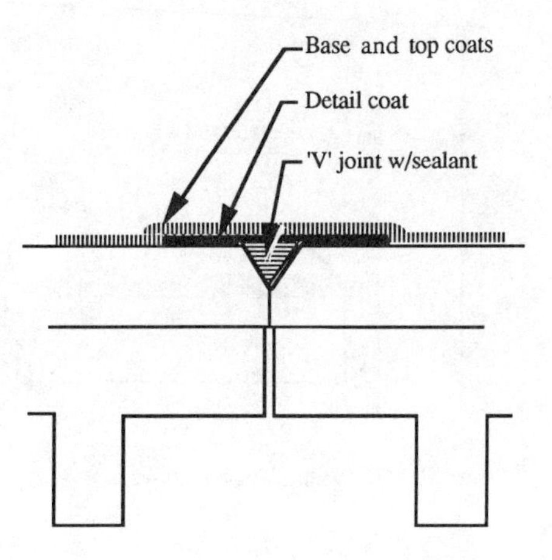

Figure 3.6 Double-T joint detailing.

Figure 3.7 Surface priming after crack preparatory work is completed. (*Courtesy of Western Group*)

film, whereas top coats range in thickness from 5–20 mil. Final coats should cure 24–72 hours before traffic is allowed on the deck, paint stripping is installed, and equipment is moved onto the deck. Approximate coverage rates for various millage requirements are shown in Table 3.19. Trowel systems are applied to considerably greater thickness than liquid-applied systems. Troweled systems range from ⅛–¼ in total thickness, depending upon the aggregate used.

Other than applications of acrylic coatings, manufacturers require primers on all substrates for improved membrane bonding to sub-

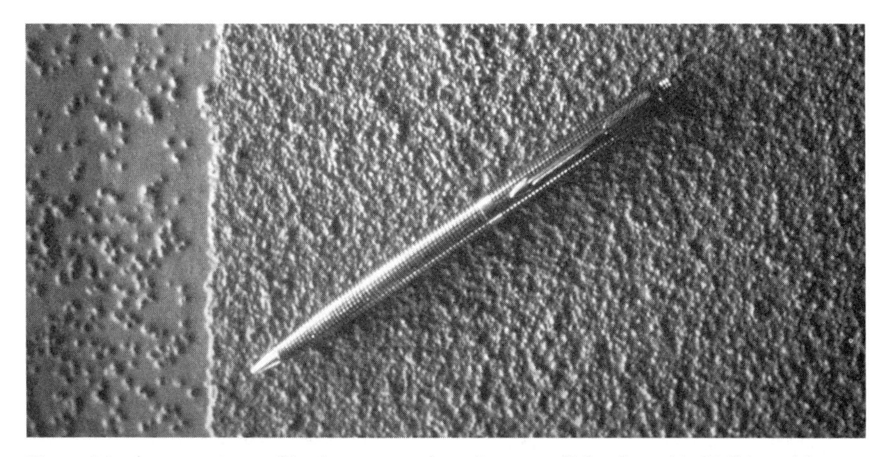

Figure 3.8 Aggregate application comparison between light duty (4–10 lb) and heavy duty (100–200 lb.) (*Courtesy of Western Group*)

TABLE 3.19 Approximate Coverage Rates for Liquid-Applied Deck Coatings

Dry millage	Coverage (ft^2/gal)
5	250
10	130
15	72
20	55
25	44
30	36
50	27

strates. Primers are supplied for various substrates, including concrete, wood, metal, tile, stone, and previously coated surfaces. Additionally, priming of aggregate or grit is required before its installation in the coating. Some primers must be allowed to dry completely (concrete); others must be coated over immediately (metal). In addition to primers, some decks may require an epoxy vapor barrier to prevent blistering from negative vapor drive.

Because of the volatile nature and composition of deck-coating materials, they should not be installed in interior enclosed spaces without adequate ventilation. Deck coatings are highly flammable, and extreme care should be used during installation and until fully cured. Deck coating requires knowledgeable, trained mechanics for applications, and manufacturer's representatives should review details and inspect work during actual progress.

Figure 3.9 demonstrates proper deck coating application, and Fig. 3.10 demonstrates backrolling procedures.

Clear Deck Sealers

Although similar to vertical surface sealers, clear horizontal sealers require a higher percentage of solids content to withstand the wearing conditions encountered at horizontal areas. Decks are subject to ponding water, road salts, oils, and pedestrian or vehicular traffic. Such inplace conditions require a solids content of 15 to 30 percent, depending on the number of application steps required. Typically, two coats are required for lower solids material and one coat for 30 percent solids material. In addition, complete substrate saturation is required rather than the spray or roller application suitable for vertical installations.

Clear wall sealers differ from elastomeric coatings much the same way that clear deck sealers differ from deck coatings. Clear deck sealers cannot bridge cracks in a substrate, whereas most deck coatings

Figure 3.9 Deck-coating application by squeegee. (*Courtesy of Western Group*)

bridge minimum cracking. Clear sealers can be applied only over concrete substrates, whereas deck coatings can be applied over metal and wood substrates. Clear sealers are penetrating systems, whereas deck coatings are surface sealers.

Properly used and detailed into other building envelope components, sealers provide the protection necessary for many deck applications. In addition, deck sealers are frequently used to protect concrete surfaces from chloride attack and other damaging substances such as acid rain, salts, oils, and carbons. By preventing water penetration, substrates are protected from the damaging effects of freeze–thaw cycles.

Unlike clear sealers for vertical applications, the chemical composition of horizontal deck sealers is limited. It includes silicone derivatives of siloxanes and silanes and clear urethane derivatives. The majority of products are siloxane based.

A sodium silicate type of penetrating sealer is available. This material reacts with the free calcium salts in concrete, bonding chemically to form a dense surface. The product is typically used as a floor hardener not as a sealer. Sodium silicates do not have properties that sufficiently repel water and the chlorides necessary for protecting concrete exposed to weathering and wear.

To ensure sealer effectiveness to repel water, test results such as ASTM C-642, C-67, or C-140 should be reviewed. Reduction of water absorption after treatment should be over 90 percent and preferably

Figure 3.10 Backrolling during deck-coating application. (*Courtesy of Western Group*)

over 95 percent. Additionally, most sealers are tested for resistance to chlorides to protect reinforcing steel and structural integrity of concrete. Tests for chloride absorption include AASHTO 259 and NCHRP 244. Effective sealers will result in reductions of 90 percent or greater.

Penetration depth is an important consideration for effective repellency and concrete substrate protection. As with vertical sealers, silanes with smaller molecular structure penetrate deepest, up to ½ in. Siloxanes penetrate to a depth of approximately ¼–⅜ in. Urethanes, containing higher solids content, penetrate substrates approximately ⅛ in.

Silicone derivative sealers react with concrete and atmospheric humidity to form a chemical reaction bonding the material to a substrate. This provides required water repellency. Substrates can be slightly damp but not saturated for effective sealer penetration. Over dense finished concrete such as steel-troweled surfaces, acid etching may be required.

Since sealers are not completely effective against water head pressures and do not bridge cracks, proper detailing for crack control, thermal and differential movement, and detailing into other envelope components must be completed. Expansion joints, flashings, and counterflashings should be installed to provide a watertight transition between various building envelope components and deck sealers.

Clear deck sealers are often chosen for application on balconies and walkways above grade, not over occupied spaces, as well as parking ga-

TABLE 3.20 Clear Deck Sealer Properties

Advantages	Disadvantages
Cost	Not completely waterproof
Ease of application	No crack-bridging capability
Penetrating applications	Not for wood or metal substrates

rage decks. In the latter, the upper deck or lower decks, which cover occupied areas, are sealed with deck coatings, while intermediate decks are sealed with clear sealers. (See Table 3.20.)

Clear Deck Sealer Application

Clear deck sealers are penetrants, and it is critical for concrete surfaces to be prepared so as to allow proper penetration and bonding. A light broom finish is best for proper penetration; smooth, densely finished concrete should be acid etched. Test applications of sealers are recommended by manufacturers to check compatibility, penetration, and effectiveness for desired results. Concrete must be completely cured, and only water curing or dissipating resin curing agents are recommended. Primers are not required with clear deck sealers.

Deck sealers should be applied directly from the manufacturer's containers in the provided solids contents and not diluted in any manner. Application is by low-pressure spray equipment, deep nap rollers, or squeegees (see Fig. 3.11). The concrete should be thoroughly saturated with the material. Brooming of material to disperse ponding collection for even distribution and penetration is then required (see Fig. 3.12). Concrete porosity will determine the amount of material required for effective treatment, typically ranging from 100–150 ft^2/gal of material.

Adjacent metal, glass, or painted surfaces of the building envelope should be protected from sealer overspray. All sealant work should be completed before sealer application to prevent joint contamination that causes disbonding of sealants.

Deck sealers are extremely toxic and should not be applied in interior or enclosed areas without adequate ventilation. Workers should be protected from direct contact with materials. Sealers are flammable and should be kept from open flame and extreme heat.

Protected Membranes

With certain designs, horizontal above-grade decks require the same waterproofing protection as below-grade areas subjected to water table conditions. At these areas, membranes are chosen much the same way as below-grade applications. These installations require a protection

Figure 3.11 Clear sealer application. (*Courtesy of Western Group*)

layer, since these materials cannot be subjected to traffic wear or direct exposure to the elements. As such, a concrete topping slab is installed over the membrane, sandwiching the membrane between two layers of concrete, therefore, the name sandwich-slab membrane.

In addition to concrete layers, other forms of protection are used, in-

Figure 3.12 Brooming of sealer application. (*Courtesy of Western Group*)

cluding wood decking, concrete pavers, natural stone pavers, and brick pavers. Protected membranes are chosen for areas subjected to wear that deck coatings are not able to withstand, for areas of excessive movement, and to prevent the need for excess maintenance. Although they cost more initially due to the protection layer and other detailing required, sandwich membranes do not require the in-place maintenance of deck coatings or sealers.

Protected membranes allow for installation of insulation over waterproof membranes and beneath the topping layer. This allows occupied areas beneath a deck to be insulated for environmental control. All below-grade waterproofing systems, with the exception of hydros clay and vapor barriers, are used for protected membranes above grade. These include cementitious, fluid-applied, and sheet-good systems, both adhering and loose laid. Additionally, hydros clay systems have been manufactured attached to sheet-good membranes, applicable for use as protected membrane installations.

Protected membranes are used for swimming pool decks over occupied areas, rooftop pedestrian decks, helicopter landing pads, parking garage floors over enclosed spaces, balconies, and walkways. Sandwich membranes should not be installed without adequate provision for drainage at the membrane elevation; this allows water on the topping slab, as well as water that penetrates the protection layer onto the waterproof membrane, to drain. If this drainage is not allowed, water will collect on a membrane leading to numerous problems including freeze–thaw damage, disbonding, cracking of topping slabs, and deterioration of insulation board and the waterproof membrane. Refer to Fig. 3.13 for an example of these drainage requirements.

Expansion or control joints should be installed in both the structural slab portion and the protection layer. Providing for expansion only at

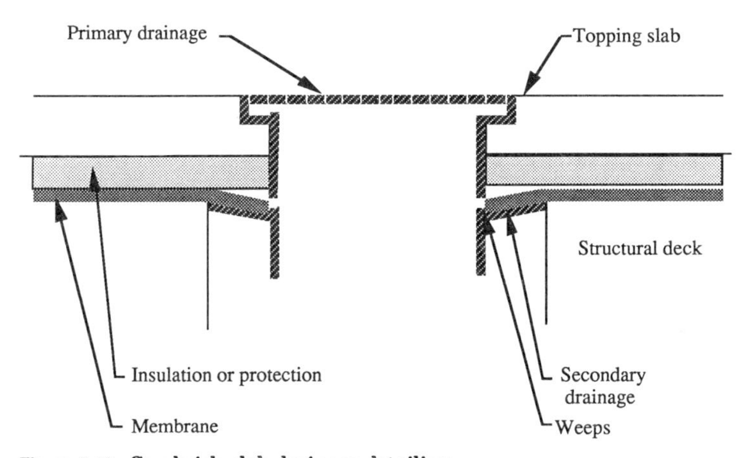

Figure 3.13 Sandwich-slab drainage detailing.

the structural portion does not allow for thermal or structural movement of the topping slab. This can cause the topping slab to crack, leading to membrane deterioration.

Membranes should be adhered only to the structural deck, not to topping layers, where unnecessary stress due to differential movement between the two layers will cause membrane failure. Refer to Fig. 3.14 for a typical expansion joint detail in split-slab construction.

Waterproof membranes should be adequately terminated into other building envelope components before applying topping and protection layers. The topping is also tied into the envelope as secondary protection. Control or expansion joints are installed along topping slab perimeters where they abut other building components to allow for adequate movement. Waterproof membranes at these locations are turned up vertically to prevent water intrusion at the protection layer elevation. Refer to Fig. 3.15 for a typical design at this location.

When pavers are installed as the protection layer, pedestals are used to protect the membrane from damage. Pedestals allow leveling of pavers to compensate for elevation deviations in pavers and structural slabs.

At areas where structural slabs are sloped for membrane drainage, pavers installed directly over the structural slab would be unlevel and pose a pedestrian hazard. Pedestals allow paver elevation to be leveled at these locations. Pedestals are manufactured to allow four different leveling applications since each paver typically intersects four pavers, each of which may require a different amount of shimming.

If wood decking is used, wood blocking should be installed over membranes so that nailing of decking into this blocking does not puncture the waterproofing system. Blocking should run with the structural drainage design so that the blocking does not prevent water draining.

Tile applications, such as quarry or glazed tile, are also used as decorative protection layers with regular setting beds and thin-set applications applied directly over membranes.

With thin-set tile installations, only cementitious or liquid-applied membrane systems are used, and protection board is eliminated. Tile is bonded directly to the waterproof membrane.

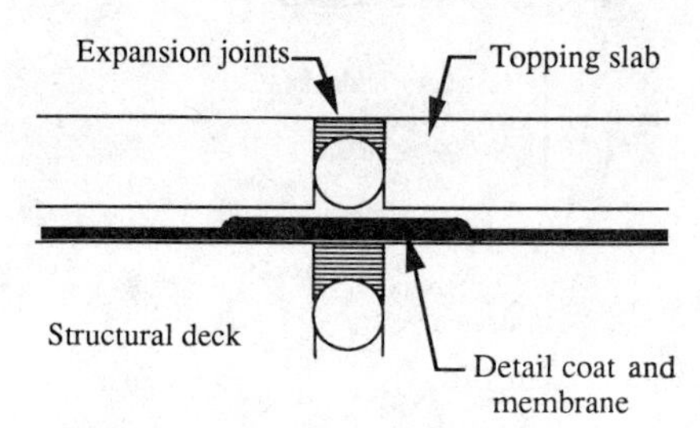

Figure 3.14 Sandwich-slab expansion joint detailing.

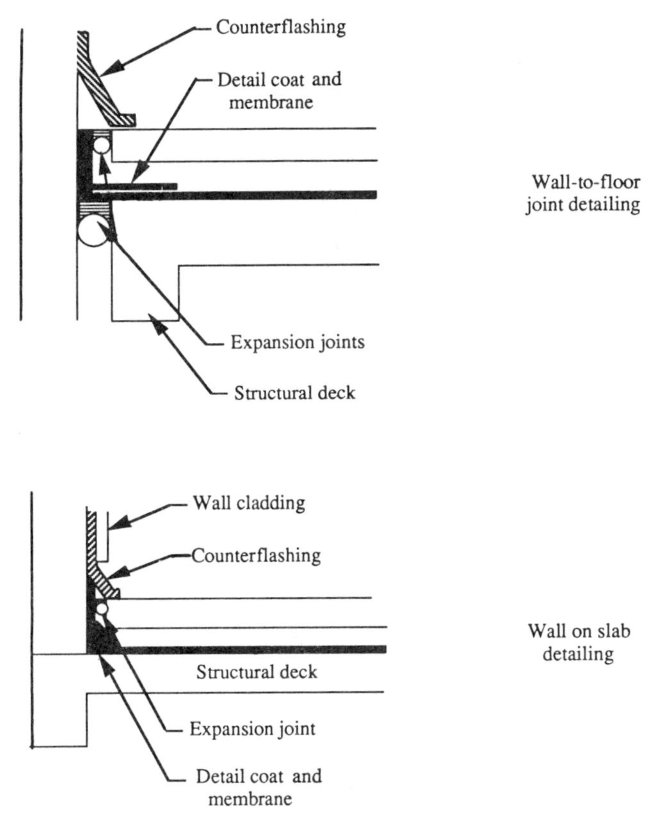

Figure 3.15 Wall-to-floor sandwich-slab detailing.

This system is used extensively for shower stall waterproofing, locker rooms, and kitchens. Tile contractors must be familiar with necessary precautions to prevent damage to the waterproof membrane.

Topping slabs must have sufficient strength for expected traffic conditions. Lightweight or insulating concrete systems of less than 3000 lb/in^2 compressive strength are not recommended. If used in planting areas, membranes should be installed continuously over a structural deck and not terminated at the planter walls and restarted in the planter. This prevents leakage through the wall system bypassing the membrane. See Fig. 3.16 for the differences in these installation methods.

Selection of a protected system should be based on the same performance criteria as those for materials used with below-grade applications. For example, cementitious systems are rigid and do not allow for structural movement. Sheet goods have thickness controlled by premanufacturing but contain seams; liquid-applied systems are seamless

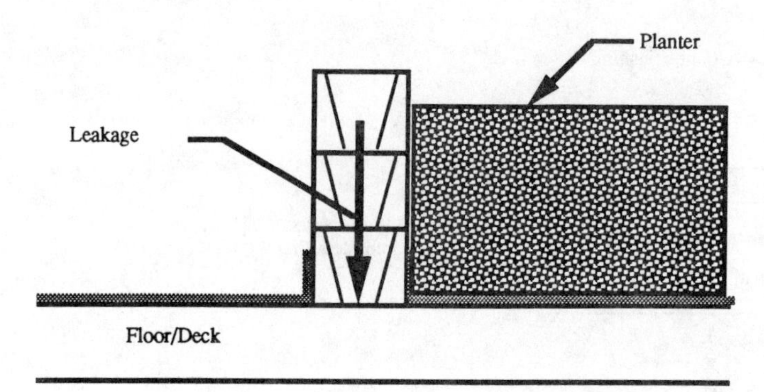

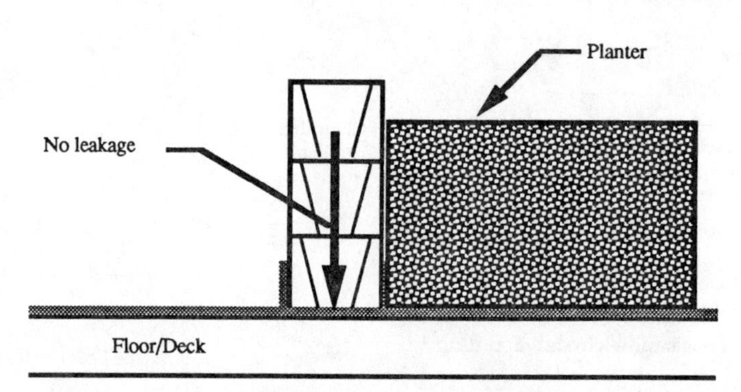

Figure 3.16 Planter detailing for split-slab membrane.

but millage must be controlled. (See Table 3.21.) Refer to Chap. 2 on below-grade systems for more specific information.

Protected Membrane Application

Guidelines for protected membranes installation are the same as for below-grade materials. Concrete surfaces must be clean and free of all lattice, dirt, and oils and must be properly cured. Most systems prohibit the use of curing agents or form-release agents.

Applications follow below-grade installation methods (see Fig. 3.17). Refer to Chap. 2 for specific installation details. With protected membrane installations, high-density insulation board can be used in place of protection board. In these applications, insulation serves two purposes—protecting the waterproofing during placement of topping

TABLE 3.21 Protected Membrane Properties

Advantages	Disadvantages
Excellent elastomeric capabilities	No access for repairs
Topping provides protection	Requires drainage at membrane level
No maintenance	No remedial applications

and providing insulation value for occupied spaces. Insulation must have sufficiently high density and compressive strength to withstand the weight of expected live and dead loads to protect cracking of toppings.

Adequate allowances should be made for movement in toppings, such as control joints or expansion joints. A water test, by complete flooding of the waterproofing system, must be completed before topping installation. This prevents unnecessary repairs to waterproofing after topping installation should leaks later be discovered.

Most sheet goods and liquid-applied systems require primers, whereas cementitious systems do not. Waterproof membranes should not be installed over lightweight concrete decks such as insulating concrete. These lightweight mixes have insufficient strength for adequate substrate usage.

Figure 3.17 Sandwich or protected-membrane application. (*Courtesy of Western Group*)

Horizontal Waterproofing Summary

Advantages and disadvantages of exposed deck coatings, sealers, and protected membranes are summarized in Table 3.22. Once chosen it is important for proper detailing or terminating into other elements of the envelope with flashings, counterflashings, control joints, or reglets for weatherproof integrity of all systems.

Roofing

Roofing is not defined as a waterproofing system, but it does form an integral part of building envelopes. Roofing is that portion of a building that prevents water intrusion into horizontal or slightly inclined elevations. Although typically exposed to the elements, roofing systems can also be internal or sandwiched between other building components.

Several waterproofing systems perform as excellent roofing systems, including fluid-applied deck coatings. Sheet and fluid systems are also used as sandwich or protected membranes as roofing components. All these systems allow roof or horizontal portions of a structure to be used for pedestrian or vehicular traffic. With such applications, insulation must be placed on the envelope's interior.

Waterproofing membranes are also used as the membrane portion of inverted roofing. These systems apply membranes directly to deck substrates with insulation and ballast over the membrane to protect it from weathering, including ultraviolet rays.

This section does not provide detail coverage of all available roofing systems that are not actual waterproofing materials; it highlights commonly used systems and their relationship and use in above-grade en-

TABLE 3.22 Characteristic Comparisons of Horizontal Waterproofing Systems

Membrane system	Advantages	Disadvantages
Deck coating	No protection or topping	Maintenance required
	Ease of repairs	Control of millage
	Crack-bridging capabilities	Limited color selections
Clear sealers	Single-step installation	No crack-bridging capability
	No protection required	Not completely waterproof
	No grit or aggregate	Highly volatile materials
Membranes	No maintenance	Protection layer required
	Crack-bridging capabilities	Interslab drainage required
	Applicable over wood and metal	No direct access for repairs

velopes are presented. In addition, Chap. 8 covers transition and termination detailing of roofing systems with other envelope components.

The most commonly used roofing systems in building construction include

- Built-up roofing
- Single-ply systems
- Modified bitumens
- Sprayed in-place urethane foam
- Metal roofing
- Protected and inverted membranes
- Deck coatings

Built-up roofing

The oldest system still commonly used today, built-up roofing (BUR), derives it name from the numerous layers of felts and bitumens applied to a substrate. Bitumens used as roofing membranes include coal tar and asphalt bitumens. By virtually adding layer upon layer, this type of system eventually covers and waterproofs a substrate and associated termination and transition details.

In the past, quality control of roofing felts used were often inadequate for in-place service conditions. As technology advanced with other systems, BUR use declined. Improvements in materials now used for felts, including fiberglass, has led to reacceptance of the built-up roofing system. Field labor safety and field quality control of actual installations do, however, present problems in most BUR applications, particularly in confined and tight areas containing intricate termination and transition detailing.

Built-up systems have poor elongation properties, with coal-tar systems being very brittle. Both asphalt and coal-tar systems do have high tensile strength. As with other sheet systems, a major disadvantage with built-up roofing is that it allows any water infiltration to travel transversely until it finds a path to interior areas. This makes leakage

TABLE 3.23 Built-Up Roofing Properties

Advantages	Disadvantages
Material cost	Material quality
High tensile strength	Low elongation properties
Multiple layers of protection	Difficult construction and safety conditions

TABLE 3.24 Single-Ply Roofing Properties

Advantages	Disadvantages
Manufactured quality control of materials	Multiple seams
Weathering durability	Termination and transition detailing
Selection of materials available	Patching and repairs

causes difficult to determine, in particular if they are being caused by rooftop equipment or associated transition detailing. (See Table 3.23.)

Single-ply roofing

Single-ply roofing systems were derived from relatively new technology for use with roofing envelope applications. Used previously for waterproofing, their adaptation to exposed conditions requires that a membrane be resistant against exposure to weathering such as ultraviolet rays.

Generic material composition of single-ply systems are as numerous as waterproofing systems previously discussed. Their applications range from fully adhered systems to loose-laid ballasted applications.

Seams continue to be a major disadvantage with any single-ply system. Seaming applications range from contact adhesives to heat welding. No seaming material or application system is, however, better than a mechanic's abilities or training in application procedures.

In addition, termination and transition detailing is extremely difficult using single-ply systems, especially those involving changes in plane such as roof protrusions. Seam installations must be carefully monitored during application to ensure installation quality. Most manufacturers have representatives who inspect installations before architectural or engineering punch list inspections. (See Table 3.24.)

Single-ply systems should be installed by manufacturer's approved applicators, and each project should receive a joint manufacturer and contractor warranty. Guarantees and warranties are discussed in Chap. 9.

Modified bitumen

Modified bitumen systems are available in a wide variety of materials and application types. These include hot-applied liquid membranes and cold-applied systems with protective top coats. Systems are available with or without fabric or felt reinforcing.

Most systems allow for seamless applications; this makes termination and transitional detailing easily detailed within the roofing instal-

TABLE 3.25 Modified Bitumen Properties

Advantages	Disadvantages
Increased performance properties	Thickness control of applications
Seamless applications	Temperature control of hot-applied systems
Termination and transition detailing	Safety concerns

lation. Modified bitumens are also used as protected membranes with inverted roofing systems. They are typically manufactured from a basic asphaltic product with added plasticizers and proprietary additives. This provides better performance characteristics as compared to basic asphaltic systems.

Generally modified bitumens are not resistant to heavy pedestrian traffic. Adequate measures must be taken to provide for walkway pads or other protection. As with single-ply systems roofing, applicators are approved and trained by material manufacturers. (See Table 3.25.)

Metal roofing

Metal roofing systems on many building envelopes are used as decorative highlights for small portions of the entire roofing area. Often metal systems such as copper domes are used for aesthetic purposes.

Complete metal roofing systems are now used regularly, especially on low-rise educational facilities and warehouse-type structures. Metal roofing systems are available in a wide variety of compositions, from copper to aluminum. They also include various manufactured pre-engineered systems.

Termination and transition detailing are difficult with metal systems, particularly when large amounts of rooftop equipment are installed. Additionally, metal roofs are not recommended for flat or minimally sloped areas that frequently occur on building envelopes.

Metal or sheet flashings are typically used for transitional detailing. This makes round protrusions and sloped areas subject to problems in

TABLE 3.26 Metal Roofing Properties

Advantages	Disadvantages
Aesthetics	Termination and transition detailing
Durability	Not for flat or minimally sloped roofs
Life-cycle weathering	Cost

detailing and possible water infiltration. If used in proper situations and expertly installed, however, certain roof systems such as copper domes will far outlast any other type of roofing installation. (See Table 3.26.)

Sprayed urethane foam roofing

Sprayed urethane foam roof systems consist of high-density urethane foam coated with an elastomeric roof coating. The foam is of sufficient density to withstand minor foot traffic. The elastomeric coating, although similar to that used for vertical envelope waterproofing, must be able to withstand ponding or standing water.

A major advantage of foam roofs is their seamless application, particularly with remedial or reroofing applications. Foam roofs can be installed over many types of existing failed or leaking roof envelopes, including built-up and single-ply roofing. Minimal preparation work is required when applying urethane foam roofs in these situations.

The urethane foam portion adds substantial insulation value to a roof, depending on foam thickness. Foams have an insulation R value of approximately 7/in of foam insulation. Urethane foam can be installed in various thicknesses and sloped to provide drainage where none currently exists. Typically, foam roofs are installed from ½–6 in thick.

After the urethane foam is installed it must be protected not from water, as it is waterproof, but from ultraviolet weathering. Foam left exposed is initially waterproof, but ultraviolet weathering will eventually degrade the foam until leakage occurs.

Thus, coating is applied to provide weathering protection for transitions and termination detailing. Coatings allow the systems to be self-flashing around roof protrusions and similar details. Foam roofs are installed in a completely seamless fashion, and their spray application makes termination and transition detailing relatively simple.

A major disadvantage with foam roofing systems is their reliance on 100 percent job-site manufacturing. Foam is supplied in two-component mixes that must be carefully mixed proportionally, heated to proper temperatures, and correctly sprayed in almost perfect weather.

TABLE 3.27 Sprayed Urethane Foam Roofing Properties

Advantages	Disadvantages
Applications for reroofing	Quality control problems caused by weather conditions
Seamless	Completely job-site manufactured system
Termination and transition detailing	Safety concerns during application

TABLE 3.28 Protected and Inverted Roofing Properties

Advantages	Disadvantages
Protection from vehicular and pedestrian damage	Difficult to repair
Protected from weathering	Drainage problems occur frequently
Provide multiple use of roof areas	Subject to insulation deterioration

Any moisture on a roof, even high humidity and condensation, will cause foam to blister as urethane foam does not permit vapor transmission. (See Table 3.27.)

Protected and inverted membranes

Sheet systems and fluid membranes used in below- and above-grade waterproofing have been successfully used for protected, sandwiched, and inverted roofing systems. These materials are identical to those previously discussed under the protected membrane section of this chapter.

Using protected membranes allows the envelope roofing portion to be used for other purposes including tennis courts and pedestrian areas. A roof area can also be used for vehicular parking when necessary structural provisions are provided.

Protrusions, particularly HVAC and electrical, are difficult to waterproof since they penetrate both structural and topping slab portions. If used, all protrusions and similar detailing should be in place and detailed before membrane installation. After the topping or protection slab is in place, protection layers should be detailed for additional protection, including movement allowances. Drainage should be provided at both topping and structural slab elevations to ensure water is shed as quickly as possible.

A major disadvantage with these systems is their difficulty in finding and repairing leakage since the membrane is inaccessible. These systems require that all applications be completely flood tested after membrane installation and before topping protection is installed to prevent unnecessary problems. (See Table 3.28.)

Deck coatings for roofing

Among the simplest and most successful but most underused roofing systems are deck-coating systems. These materials are used primarily for waterproofing pedestrian and vehicular decks as previously discussed in this chapter. Deck coatings applied as roofing systems provide seamless applications, including terminations and transitions,

TABLE 3.29 Deck Coatings Used for Roofing Properties

Advantages	Disadvantages
Seamless applications	Insulation must be placed on the underside of the deck
Seamless termination and transition detailing	Control of millage thickness
Allows roof areas to be used for different purposes	Subject to blistering by negative moisture drive

and are completely self-flashing. They can be applied over wood, metal, and concrete substrates.

The only major disadvantage in roofing application of deck coating is that insulation must be installed to the underside of the exposed envelope portion. If this is possible, deck coatings provide numerous benefits when installed as roofing systems.

Since deck coatings adhere completely to a substrate, water cannot transverse longitudinally beneath the system. Therefore, should leakage occur, it will be directly where a membrane has failed and easily determinable. These membranes are resistant to pedestrian and vehicular traffic, providing resistance to abuse that most other roofing systems cannot.

Repairs are easily completed by properly repairing and recoating an affected area. Any equipment changes, roof penetrations, and so forth, can be made after the initial roof installation. These repaired areas also become seamless with the original application.

Deck coatings are applied directly from manufacturer's containers in a liquid. This application eliminates the need to heat materials, seam, and use spray application, which are required with other roofing systems. By providing the simplest installation procedure, deck coatings eliminate most human error and result in successful roofing systems for building envelopes. (See Table 3.29.)

Roofing Installation

Mechanical equipment, plumbing stacks, and electrical penetrations are often poorly detailed, presenting almost impossible conditions in which to install roofing materials. The difficult areas include inaccessible places beneath mechanical equipment and electrical conduit protrusions too close to adjacent equipment to properly flash roof transition materials.

By limiting the number of roof penetrations, providing areas large enough to install transition detailing, and ensuring minimum heights above roof for termination, detailing will limit common 90 percent leakage problems. All equipment should be placed on concrete curbs a minimum of 8 in above roofing materials. Wood used for curbs can rot and eventually damage the roof. This minimal height provides sufficient areas to transition and terminate roofing materials properly into equipment that becomes part of the building envelope.

Curbs should be placed so as to complement roof drainage and not block it. Roofing membranes should extend both under the curb and over it, completely enveloping it to prevent leakage.

Any conduits or drains running to and from equipment should be raised off the roof so as not to prevent drainage and damage to roof membranes. Any rooftop mounted equipment such as balustrades, signs, and window washing equipment should be placed on curbs. Equipment fasteners used at curb detailing, as well as the equipment itself, should be waterproofed to prevent transition and termination water infiltration.

Roof penetrations and all protrusions, such as electrical conduits, should be kept in as absolute minimum of groupings as possible. Roofs should not be used as penthouse areas for electrical and mechanical equipment nor storage areas for excess equipment. Too often equipment is placed on a roof in groupings that make maintenance and drainage impossible. Further, any equipment added after roofing completion should be reviewed by the material manufacturer and roofing contractor to ensure that warranties are not affected by the installation.

Roofs should be tested for adequate drainage before membrane installation. Once rooftop equipment is installed, drainage should be checked and adjustments made where necessary. After roofing is installed, it is too late to repair areas of ponded water. Roof drains must be placed at the lowest elevations of the roof and not be obstructed by roof to equipment.

All related roof envelope equipment should be tested for watertightness after installation. Roofing envelope portions are often damaged by equipment that allows water or condensation to bypass roof membranes and to enter directly into interior areas.

Sealants should not be used excessively as termination or transition detailing anywhere within the roofing envelope. Sealants typically have a much shorter life cycle than roofing membranes. Sealants then become a maintenance problem that when not properly attended to create leakage.

Roofing Summary

As with a complete envelope, it is not typically roofing material or systems that directly cause water infiltration; it is the 1 percent of a roofing envelope portion. This 1 percent includes termination and transition details including flashings, protrusions, and mechanical supports that typically occur within a roofing application.

Roofing envelope installation often involves more subcontractors and trades than any other building envelope portion; these people range from sheet metal mechanics to window washing equipment installers. This extreme multiple discipline involvement requires utmost care in detailing and installing termination and transition details, not only between various rooftop components but transitions between roofing and other building envelope components of the envelope. These transition details are covered fully in Chap. 8.

Vapor Barriers

Vapor barriers are used in above-grade construction to prevent moisture vapor transmission between interior and exterior areas. In winter conditions, warm moist interior air is drawn outward to the drier outside air by the difference in vapor pressures (negative vapor drive). In summer, moisture vapor travels from moist and warm outside air to cool and dry interior areas (positive vapor drive).

Vapor barriers or retarders are not waterproofing materials but are used as part of wall assemblies to prevent vapor transmission and allow this vapor to condensate into liquid form. Vapor barriers are most useful in hot tropical areas where vapor transmission into air-conditioned areas can be so severe that mold and mildew frequently form on exterior walls. This problem is often mistaken for water leakage or infiltration when it is not. Attempts to repair, including applying breathable coatings to an envelope (e.g., elastomeric coatings) will not solve this problem.

Should a nonbreathable coating be applied to an envelope under these conditions, however, coating blistering and disbonding will occur when negative vapor drives occur. This requires that vapor retarders or barriers be applied to the interior or warm side of insulated areas. In tropical areas the barrier is placed on exterior sides to prevent condensation or vapor from wetting insulation caused by positive drive. In most areas, barriers are placed on interior sides of insulation due to the predominance of negative vapor drive.

Vapor barriers are commonly available in polyethylene sheets or aluminum foil sheets on laminated reinforced paper. Sheets must be applied with seams lapped and sealed to prevent breaks in the barrier.

A vapor barrier's performance is measured in perms (permeability). This is the measure of vapor transmitting through a particular envelope material or component. Materials such as masonry block have high permeability, whereas polyethylene materials have very low permeability. Glass is an example of a barrier. Moisture collects and condenses on glass because it cannot pass through the glass.

Although vapor barriers are not used as waterproofing systems, they can affect the selection and use of waterproofing materials for use on an envelope. If negative vapor drive is possible (winter conditions), it is necessary for permeable waterproof materials to be used to allow this moisture to pass without damaging the waterproofing material by blistering or delaminating.

Chapter

4

Sealants

Sealants are not only the most widely used waterproofing materials, but also the most incorrectly used. Although sealants are a relatively minor cost item, they constitute a major function in a building's life cycle. Applied from below-grade to roof areas, and used as components of complete waterproofing systems and for detailing junctures and terminations, sealants act as direct waterproofing barriers. As such, sealants are important in constructing successful watertight building envelopes. Sealants are also used to prevent air from infiltrating in or out of a building. Sealants thus have a dual weatherproofing role, with waterproofing as the primary role and environmental control as the secondary role.

Practically every building's exterior skin requires sealants for weathertightness. Junctures of dissimilar materials or joints installed to allow for structural or thermal movement require sealants to maintain envelope effectiveness. Below grade, sealants are used for sealing expansion joints, junctures, or terminations of waterproofing compounds and protrusions. Above-grade applications include sealing joints between changes in building facade materials, window and door perimeters, and expansion and control joints. Sealants are also used to detail numerous joints, including flashings and copings that act as termination or transition details.

Since sealants are a minor portion of overall construction scope, they receive a comparable amount of effort in their design and installation. Yet, because they are a first-line defense against water infiltration, sealant failures can cause an unequal proportion of problems and resulting damage (Fig. 4.1).

Technologically, sealants have advanced dramatically from the white stuff in a tube, and a clear differential should be made between caulk-

Figure 4.1 Failure of sealant joint by excess movement at joints. (*Courtesy of Western Group*)

ing, sealants, and glazing materials. *Caulking* refers to interior applications, to products manufactured for interior use and installed by paint contractors. These materials are usually painted after installation. Caulking is installed as a filler between dissimilar materials in an interior controlled environment not subject to thermal or structural movement. Therefore, caulking does not require the performance materials that exterior high-movement joints do. *Sealants* are exterior applications using high-performance materials (e.g., silicones), which are typically colored rather than painted and are applied by waterproofing contractors.

Sealants are also differentiated from *glazing materials,* which are considerably higher in tensile strength. This higher tensile strength produces lower elongation capabilities than sealants or caulking possess. Glazing materials are used in construction of window panels or curtain walls where higher strength is more important than movement capability. This strength (tensile) is referred to as the modulus of elasticity.

Successful installation of sealant depends on several steps, including

- Joint design
- Material selection
- Substrate preparation
- Joint preparation and installation

Each step is critical for sealant systems to perform successfully for extended periods of time. Better sealant materials will perform for 5–10 years. But because of improper design, incorrect material choices, poor installation, or a combination of these factors, sealant joints rarely function within these time parameters.

Joint Design

Joint design failures are often attributable to improper spacing and sizing of joints. Joints are frequently arranged for aesthetic purposes, and actual calculations to determine optimum number and spacing of joints are overlooked. Precast and prefabricated panel joints are often determined by panel size of an individual precast manufacturer rather then by sound joint design.

Even if joint size requirements are actually determined, far too often panel erectors are primarily concerned with the installation of panels, with joints being used to absorb installation tolerance during erection. Often these joints end up varying greatly in width from those originally intended, with no procedures followed for maintaining proper tolerance in joint width.

Joint type

The first step in proper joint design is to determine areas and joint locations required within a building envelope. Areas of change in materials (e.g., from brick to concrete), of changes in plane, of differential movement potential (e.g., sprandrel beams meeting columns), of protrusions (e.g., plumbing and ventilation equipment), and of thermal movement all must be studied to determine location requirements for joints.

Once this study is complete, design calculations must be completed to determine the type, spacing, and size of joints. Joint types include

- Expansion joints
- Control joints
- Isolation joints
- Detailing joints

Expansion joints. Expansion joints allow for movement in a structure or material that is caused by thermal expansion or contraction and other inducements such as wind loading or water absorption. Expansion joints are active dynamic movement joints that continue to move by expansion or contraction. See Fig. 4.2 for a typical masonry expansion joint.

Control joints. Control joints allow for expected cracking due to settling, curing, or separation in building materials after installation. Interior control joints, including concrete slab control joints, are typically nonmoving joints and are placed and sized for expected cracking or

Figure 4.2 Expansion joint at masonry envelope ready for sealant application.

shrinkage only. Exterior control joints, such as brick paver joints, provide for settling as well as movement, the latter due to vehicular or pedestrian loading and expected thermal movement. These joints require more design work than interior joints, as they will become dynamic moving joints.

Masonry and mortar shrinkage after placement and curing requires that control joints be placed at appropriate locations. These joints allow for shrinkage and settlement to occur without affecting an envelope's performance. Control joint locations should include

- Areas of change in wall height
- Junctures or transitions at columns or other structural construction
- Wall intersections or changes in plane
- Areas where large openings occur in the envelope, such as above and below window openings

Isolation joints. Junctures at changes in materials require isolation joints to allow for any differential movement between two different materials. Window frame perimeters abutting facade materials and like areas of change in structural components (e.g., sprandrel beam meeting brick facing material) require sealant joints because of differential movement.

Detailing joints. Detailing joints are required as a component or part of complete waterproofing systems. They are used to impart watertightness at building details such as pipe penetrations and changes in plane before the application of primary waterproofing materials.

Spacing and sizing joints

Once the appropriate types of joints are determined, calculations are necessary to determine proper spacing and sizing of the required joint opening. Following are established guidelines used frequently in the industry; note that these are not meant to replace actual engineering calculations.

Basic rules for joint design include

1. Joint size no smaller than ¼ in
2. Joint size no larger than 1 in
3. Joint opening a minimum of four times anticipated movement

Basic rules for sealant design include

1. Material thickness no less than ¼ in
2. Joints up to ½ in wide; depth of material is equal to width of material
3. Joints wider than ½ in; depth of material is one-half the width
4. Maximum recommended width is 1 in
5. Maximum depth is ½ in

The *number and spacing of joints* are determined by

1. Anticipated substrate movement, determined by coefficient of expansion
2. Length of substrate material span
3. Joint width

Design for movement is usually based on a temperature differential of 150°F. This is movement occurring in a selected material in a change of temperature of 150°F. Coefficients of thermal expansion are usually expressed as inch per inch per degree Fahrenheit. To determine expected movement and resulting joint size, the coefficient of linear expansion is multiplied by temperature range, span length of material, and appropriate safety factor (usually at least a factor of 4).

The following is a typical calculation for joint design. As an example using 10-ft concrete precast panels with a coefficient of 0.000007 in/in/°F, the following would be the recommended joint design width:

$$150°\text{F} \times 0.000007\ \text{in/in/°F} \times 10\text{ft} \times 12\ \text{in/ft} \times 4\ \text{(safety factor)} = 0.504 = ½\ \text{in}$$

In determining the joint size necessary for moving joints located at different material intersections, materials with highest coefficients of

TABLE 4.1 Coefficients of Thermal Expansion for Common Envelope Materials

Material	Coefficient of thermal expansion, in/in/°F
Aluminum	0.000013
Concrete	0.000008 to 0.000005
Granite	0.000005
Limestone	0.000005
Marble	0.000007
Masonry	0.000004 to 0.000003
Plate glass	0.000005
Structural steel	0.000007
Wood	0.000002 to 0.000003

expansion are used in calculations. However, if a material with lower movement coefficient is spanning a greater width, these data may present a larger joint size. Therefore, it is necessary to calculate all possible combinations to determine the largest joint size necessary. Table 4.1 summarizes coefficients of thermal expansion for several common building materials.

Backing materials

It is imperative that sealant materials be allowed to expand to their maximum capability without exerting unnecessary stress at the adhered substrate surface area. Thick beads of sealant are more difficult to elongate (a thick rubber band is harder to stretch than a thin band), which places more stress on the adhered area of substrate. If this stress exceeds a sealant's capability, adhesive failure will occur.

To prevent failure, a backing material is inserted into joints to provide a large adhered contact area, at two sides of a joint, with a thin bead of sealant. This is shown in Fig. 4.3. This backing material, or backer rod as it is commonly referred to, is of major importance in joint design and application. Besides ensuring proper joint design, the backer rod allows applicators to monitor proper depth of material installation. It also provides a surface against which uncured sealant material can be tooled so as to force it against both sides of a joint for proper adhesion.

Sealants do not adhere to the backer rod, only to joint sides. Three-sided adhesion places too much stress on sealant material in movement (elongation), causing tears that result in cohesive failures.

Two types of backer are commonly used: open-cell polyurethane and closed-cell polyurethane. Both serve the same purpose, but sealant ma-

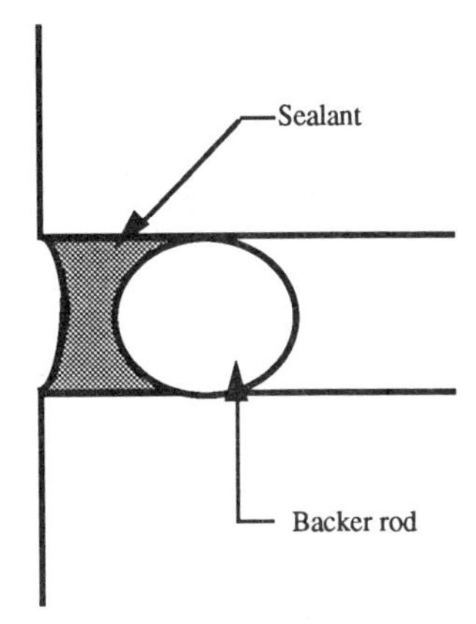

Figure 4.3 Backing detailing in sealant joints.

terial selection limits which backer material is used. Open-cell rod is compatible with most sealants as long as the backer remains dry (its open-cell structure allows for the absorption of moisture). If contaminated with moisture, it will prevent sealant curing, as moisture will remulsify sealants. Closed-cell rod is not susceptible to moisture, but it is not used with air-cured material since it prevents curing the unexposed joint side.

Backing materials are held in place by compressing the material between joint sides. This requires backing to be at least 25 percent larger than the joint width. In joints not sufficiently deep for installing a backer rod, a bond breaker tape should be used. This type of tape prevents sealants from adhering to backs of a joint and preventing three-sided adhesion. This is typical of slab or horizontal joints. The tape is also used for window perimeter joints, where sufficient space for backing materials is often not provided.

Bond breaker tape must be made of a material that will not bond to sealants. Masking tape or duct tape is not acceptable. Backing tape is typically of polyethylene composition, 12–20 mil thick. It is extremely important that the tape not be so narrow that the sealant is allowed to adhere on the backside of a joint. Tape should also not be so wide that it turns up joint sides, preventing proper adhesion. A proper fit can be accomplished by cutting installed tape that is slightly larger than joint width along the joint edges with a razor knife and removing excess tape.

Joint detailing

Sealant materials are limited in movement capability by the dimension of their narrowest width. This leads to failure of joints placed in a V-shape at window frames and glass-to-metal or metal-to-masonry junctures. These joints are extremely limited in movement capability at the base or bottom of the V-formation, which causes cohesive tears, severely limiting elongation or leading to material failure.

Many construction cladding materials used today are permeable to water. Concrete, precast concrete, masonry blocks, and brick all allow water or vapor to enter directly through the building cladding and bypass sealant joints. In addition, water enters through substrate cracks, defective mortar joints, and other envelope voids. In many instances, considering that field construction is not a perfect science, sealant joints may simulate two sponges sealed together.

When water bypasses a joint through a substrate it travels transversely on a path of least resistance. Water then collects at backs of substrate breaks or joints (usually where a sealant joint is installed), finding a path into the interior drawn by the difference in air pressure between interior and exterior. This leakage often appears as joint leakage, when in fact it is due to substrate permeability.

Therefore, it is often prudent to double seal exterior joints. The secondary joint effectively seals interior areas from water intrusion, bypassing initial sealant joints. If accessible, second joints are sealed from the exterior, but they can be sealed from inside the structure. In both cases, joint design should include allowances for drainage of moisture that passes the first joint, back to the exterior, by installing flashing and weeps. Double seal designs should not include materials that are sensitive to negative moisture drive, which is present in these applications.

Double sealing has several advantages beyond those derived from waterproofing. This joint design stabilizes air pressure in the space between sealant beads, thus eliminating positive vapor transmission into a building by air pressure. The interior bead also stops vapor inside a building from entering into cladding where it may condense and cause damage, such as spalling or corrosion, to building components. This double-sealant installation also serves as an energy conservation method by effectively eliminating uneven air pressures that cause airflow into or out of a structure.

Design of inner sealant beads is controlled by design of exterior joints. Since both sides of a joint movement will be equal, the same material should be used on both. Using sealants with low movement capability on interior sides leads to ineffectiveness in preventing water and air transmissions.

If single-sealed joints are to be used on an envelope, substrates that form the joints should be constructed or manufactured to shed water quickly from the joint and envelope. They should also be designed to prevent water from traveling laterally across the joints.

Figure 4.4 shows several primary envelope barrier designs that complement joint effectiveness. These allow joint sealants to be the second-

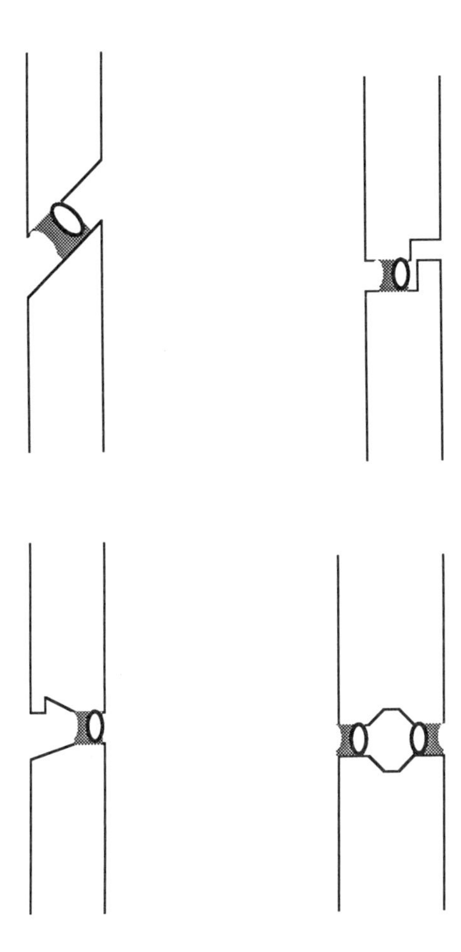

Figure 4.4 Envelope-joint construction for effective sealing.

ary means of protection against water infiltration. These designs also provide secondary protection against direct water infiltration should a sealant exhibit minor disbonding or adhesion problems along the joint. Never should an envelope joint be designed that allows water to stand or collect on the sealant material.

Material Selection

In addition to the elongation properties necessary for the expected movement the most important sealant properties are

- Adhesion strength
- Cohesion strength
- Elasticity

- Expected life
- Modulus

Additional desirable features or characteristics include color, availability, paintability, substrate compatibility, ultraviolet resistance, and presence of one or two component materials.

Elongation

Elongation is the ability of sealants to increase in length then return to their original size. Limits of elongation are expressed as a percentage of original size. A material with a 200 percent elongation ability is, therefore, capable of stretching to double its original size without splitting or tearing. Since this is an ultimate measure of failure, joints are not designed to perform to this limit of elongation, rather to a portion of this capability including a safety factor consideration. A joint stretched to its limit will not return to its original shape or size. Thus a joint will cease to function properly if elongated to its maximum elongation capability.

Modulus of elasticity

The modulus of elasticity is the ratio of stress to strain and is measured as tensile strength, expressed as a given percentage of elongation in pounds per square inch. Modulus has a direct effect on elongation or movement capability. Low-modulus (tensile strength under 60 lb/in^2) materials have a higher ability to stretch than high-modulus sealants. High tensile strength results in lower movement capability. More simply, soft materials are more easily stretched than harder materials. Low-modulus sealants with high-elongation factors are required in high-movement joints.

Elasticity

Elasticity and recovery properties are measures of a sealant's ability to return to its original shape and size after being compressed or elongated. As with elongation, elasticity is measured as a percentage of its original length. In high-movement joints, a sealant with sufficient recovery ability is mandatory. A sealant that does not continually return to its original shape after movement will eventually fail.

Adhesive strength

Adhesive strength is the ability of sealants to bond to a particular substrate, including adhesion during substrate movement. Since materials differ substantially in their adhesive strength to particular substrate, manufacturers should be consulted for adhesion test samples on proposed substrates.

Cohesive strength

Cohesive strength is the ability of a material's molecular structure to stay together internally during movement. Cohesive strength has a direct bearing on elongation ability.

Shore hardness

Shore hardness is resistance to impact, measured by a durometer gage. This property becomes important in choosing sealants subject to punctures or traffic, such as horizontal paver joints. A hardness of 25 is similar to a soft eraser; a hardness of 90 is equivalent to a rubber mallet.

Material testing

All of the properties discussed must perform in unison for materials to function as necessary during joint life cycling. Weathering, ultraviolet resistance, amount of movement, and temperature change all affect sealant durability.

A sealant can exhibit excellent properties of elongation in a manufacturer's laboratory, but if it is not resistant to the elements the material soon fails. Many tests are available for comparison of different materials as well as different manufacturers.

Unbiased testing is completed by the National Bureau of Standards (federal specifications) and the American Society for Testing and Materials. Tests presently used as standards include

Federal specification TT-S-227 for two-component sealants

Federal specification TT-S-00230C for one-component sealants

Federal specification TT-S-001543 for silicone sealants

ASTM tests included in volume 4.07 for building seals

These testing agencies provide tests from which to compare materials between various manufacturers. For example, an ASTM test of accelerated aging produces a comparison of a particular sealant's ability to withstand weathering to those of other manufacturers. If all such results are produced by the same testing guidelines, they can be used as basis for comparisons. Although a generic type of sealant may vary somewhat in properties from manufacturer to manufacturer, the basic properties are inherent in a particular class of sealant.

Materials

The numerous materials used as sealants exhibit a wide range of properties. In choosing a sealant, properties should be matched to expected conditions of a particular installation. The most common materials available and used for sealing joints in building construction include

- Acrylic
- Butyl
- Latex
- Polysulfide
- Polyurethane
- Silicones
- Precompressed foam
- Preformed derivatives

Once the joint design is completed, a material with required properties must be chosen. Typical properties of each class of sealant are summarized in Table 4.2. Preformed seals are considered in Chap. 5 on expansion joints.

Acrylics

Acrylic-based sealants are factory mixed one-component materials polymerized from acrylic acid. These are not used on joints subject to high movement because of their relatively low-movement capability. They are frequently used in remedial applications with acrylic-based waterproof coatings. Acrylic materials are available in brushable or trowel grades for use in preparing cracks in substrates before waterproof coating application. They are used in small movement joints such as doors and window perimeters, thresholds, and equipment penetrations.

Acrylic-based sealants do not require primers and have minimal surface preparation. Their general ease of application is offset by low performance characteristics. These materials are not recommended in continually submerged joints or joints subject to vehicular or foot traffic. (See Table 4.3.)

Butyl

Butyl sealants are produced by copolymerization of isobutylene and isoprene rubbers. Butyls are some of the oldest derivatives to be used for sealant materials. However, technological advancements in better performing sealants have now limited their use to glazing window perimeters or curtain walls with minimal movement.

TABLE 4.2 Comparison of Common Sealant Properties

Property	Acrylic	Butyl	Latex	Poly-sulfide	Poly-urethane	Sili-cone	Precom-pressed foam
Maximum joint movement capa-bility, %	7	5	7	25	25	50	25
Weathering resistance	Good, excel-lent	Excel-lent	Fair	Good	Excel-lent, Good	Excel-lent	Excel-lent
Recovery, %	25	Poor	75	80	90	100	100
Adhesion	Good	Excel-lent	Fair	Good	Good	Excel-lent	Excel-lent
Joint design (number of times expected movement)	12	20	12	6	4	4	*
Shrinkage, %	12	18	20	10	5	3	n/a
Tack-free time (hours)	72	24	1	72	72	3	n/a
Water immersion	No	No	No	Yes	Some	No	No
Paintable	Yes	Yes	Yes	No	No	No	No
Primer required	No	No	Some cases	Metal masonry	Hori-zontal masonry	Metal natural stone	No
Ultimate elongation, %	Low	Low	450	1000	700	1600	Very low
Horizontal joints	No	No	No	Yes	Yes	No	Yes
Modulus of elasticity, lb/in^2	40	25	18	30	35	30	25

*Best in compression mode.

Although butyls have low-movement and recovery characteristics, they have excellent adhesion performance. They bond tenaciously to most substrates and have excellent weathering characteristics. Butyls should not be used on water-immersed joints or joints subject to traffic.

Butyl sealants are used in metal curtain wall construction because of their ability to function in very thin applications. As long as movement is within the capability of a butyl, materials will function properly in metal wall construction splice joints.

Butyls are relatively easy to install, available in one-component

TABLE 4.3 Acrylic Sealant Properties

Advantages	Disadvantages
No primers required	Long cure stage
Good UV resistance	Low-movement capability
Minimal surface preparation	Poor impact resistance

packaging, and easily gunable or workable. They require no priming and are paintable. (See Table 4.4.)

Latex

Latex sealants are typically acrylic emulsions or polyvinyl acetate derivatives. Latex materials have very limited usage for exterior applications. They are typically used for interior applications when a fast cure time is desired for painting. Latex sealants have an initial set of tack-free time of less than 1 hour, fastest of all sealant materials.

Latex materials have very low movement capability, high shrinkage rates, and only fair weathering and adhesion properties. Their exterior use is limited to window or door perimeters where it is desired that the sealant match the frame color opening. Latex materials should not be used in areas subject to water immersion or traffic. (See Table 4.5.)

Polysulfides

Polysulfide materials are produced from synthetic polymers of polysulfide rubbers. Polysulfides make excellent performing sealants for most joint use. However, urethanes and silicones frequently have become specified and used due to their excellent recovery ability and joint movement capability.

Polysulfides withstand an average of 16–20 percent joint movement with a joint design of six times anticipated movement versus a joint movement of 25 percent for urethanes and joint design of four times anticipated movement.

As with other types of better sealants, polysulfides exceed the movement capabilities of paints and therefore should not be painted. They

TABLE 4.4 Butyl Sealant Properties

Advantages	Disadvantages
No primers required	Low-movement capability
Excellent weathering	High shrinkage rate
Little surface preparation	Poor recovery

TABLE 4.5 Latex Sealant Properties

Advantages	Disadvantages
Fast cure stage	Low-movement capability
Paintable	High shrinkage rate
One component	Poor weathering

are, however, manufactured in both one- and two-component packaging in a wide range of colors. With two-component materials, a color additive is blended in during mixing. Color charts are provided by the manufacturer.

Polysulfides are acceptable for a wide range of applications, including curtain wall joints, precast panels, and poured-in-place concrete. Polysulfides require primers on all substrates, and preparation is critical to allow successful adhesion and movement capabilities of installed materials.

Manufacturers usually produce two types of primers—one for masonry, concrete, and stone, and another for glazing, glass, and aluminum work. In a precast panel to window frame perimeter joint, two different types of primer on each side of the joint would be required.

If properly prepared and installed, polysulfides will function in constantly immersed joints. Of all commercially available sealants, polysulfides are best suited for total immersion joints. This includes swimming pools, water and wastewater treatment structures, fountains, and water containment ponds. Typically, two-component materials are recommended for these types of joint installations.

Polysulfides should not be installed in joints that might have bituminous residue or contamination, such as premold joint filler (e.g., concrete sidewalk joints). Polysulfides should also not be applied over oil- or solvent-based joint sealants. Joint preparation for resealing joints containing asphalt or oil-based products is especially critical if polysulfides are to be used. Sandblasting or grinding of joints to remove all residues is necessary before application of polysulfide materials.

Polysulfides are manufactured in grades for horizontal joints subject to foot or limited vehicular traffic. These materials are self-leveling and ideal for plaza and parking deck joints. (See Table 4.6.)

Polyurethane

Urethane sealants are polymers produced by chemical reactions formed by mixing di-isocynate with a hydroxyl. Many urethanes are moisture-cured materials reacting to moisture in atmospheric conditions to promote curing. Other two-component urethanes are chemically curing mixtures. Their compatibility with most substrates and

TABLE 4.6 Polysulfide Sealant Properties

Advantages	Disadvantages
Immersion applications	Primers required
Good UV resistance	Low-movement capability
Horizontal applications	Low recovery rates

waterproofing materials has made them a commonly used sealant in waterproofing applications.

Formulations range from one-component self-leveling materials in a pourable grade for horizontal joints in plaza decks to two-component nonsagging materials used for vertical expansion joints. Some urethanes are manufactured to meet USDA requirements for use in food processing plants. As with polysulfides, polyurethanes are available in a wide range of colors. Two-component mixes add coloring to the activator portion that is mixed with base material.

Polyurethanes are available for a wide range of applications, including precast concrete panels, expansion and control joints, horizontal joints, flashing, and coping joints. Urethane sealants are not recommended for continual immersion situations.

Urethane has excellent adhesion to most substrates, including limestone and granite. In most cases a primer is not required. However, manufacturer's data should be reviewed for uses requiring primers. These include horizontal joints, metals, and extremely smooth substrates such as marble.

Two-component urethanes are low-modulus sealants and have high joint movement capability averaging 25 percent, with joint design limitation of four times the expected movement. Since urethanes exceed the movement capabilities of paint, they should not be painted over, because alligatoring of the paint surface will occur. Coloring should be achieved by using standard manufacturer colors.

Urethanes have excellent recovery capability, 90 percent or more, and possess excellent weathering and aging characteristics. Since urethanes are extremely moisture sensitive during curing, a closed-cell backer rod should be used. However, with one-component urethane sealants an open-cell backing material is acceptable.

Polyurethanes cannot be used in joints containing a polysulfide sealant or residue. These joints must be cleaned by grinding or other mechanical means to remove any trace of sulfides. Urethane sealants also should not be used in glazing applications of high-performance glass, plastics, or acrylics. Joints contaminated with asphalts, tar, or form release agents must be cleaned before sealing work.

Polyurethane's compatibility with most substrates, excellent move-

TABLE 4.7 Polyurethane Sealant Properties

Advantages	Disadvantages
Good elongation capability	Moisture sensitive
Excellent recovery rates	Unpaintable
Horizontal applications	Require some priming

ment and recovery capability, and good weathering characteristics have allowed their widespread use in waterproofing applications both above and below grade. Their ability to withstand vehicular traffic and compatibility with urethane deck coatings leads to their extensive use in parking deck applications. (See Table 4.7.)

Silicones

Silicone sealants are derivatives of silicone polymers produced by combining silicon, oxygen, and organic materials. Silicones have extremely high thermal stability and are used as abrasives, lubricants, paints, coatings, and synthetic rubbers. Silicones are available in a wide range of compositions that are extremely effective in high-movement joints, including precast panels and expansion joints. When used properly, silicone sealants provide excellent movement capability, as much as 50 percent, and adhesion and recovery properties after movement.

Silicone sealants cannot be used for below-grade applications, horizontal applications subject to vehicular traffic, and water immersion joints. It is extremely important not to install silicone materials over materials that might bleed through a silicone. This includes oil, solvents, or plasticizers, which will cause staining and possible silicone failure.

Uncured silicone must not encounter nonabradable substrates such as metal, polished granite, or marble. The uncured sealant can leave a residue that stains or changes the substrate appearance. This is also true for primers used with silicone sealants. Masking adjacent surfaces is necessary to protect against damage.

Silicones contaminate all surfaces or substrates they encounter. This makes it virtually impossible to seal over silicone residue with other materials such as urethanes. Abrasive methods are the only acceptable methods for removing silicone from a substrate before resealing it with another product. Most substrates do not require primers for silicone applications; however, natural stone materials such as limestone and marble will require primers.

Most silicone is produced in one-component packaging, although two-component products are available. Silicones have excellent adhe-

TABLE 4.8 Silicone Sealant Properties

Advantages	Disadvantages
High-movement capability	No submersion applications
Excellent adhesion	Possible staining
Excellent recovery rates	No below-grade uses

sion to almost all building products including wood, ceramic, aluminum, and natural stones. Silicones may be used in curtain wall joints, precast panels of concrete, marble, or limestone, and expansion and control joints.

Silicone materials exceed the movement capability of paints, and as most paints will not adhere to silicones, they should not be painted over. Most silicones are now produced in a wide range of colors; in addition, special color blending is available by the manufacturer. Both open-cell and closed-cell backing materials can be used with silicone joints.

Silicones have excellent recovery capabilities, usually up to 100 percent. They have very little initial cure shrinkage, 3 percent, and a tack-free time of only 1–3 hours. High tensile strength silicones with lower movement capabilities are typically used in glazing applications. (See Table 4.8.)

Precompressed foam sealant

Foam sealants are manufactured by impregnating open-cell polyurethane foam with chemical sealant containing neoprene rubbers. An adhesive is applied to one side and covered with a release paper. The foam is then compressed and supplied in rolls with various widths up to 12 in.

Foam sealants are applied by unrolling the foam, removing release paper exposing the adhesive, and installing into a joint. The foam then swells and expands to fit tightly against both joint sides, allowing for any irregularities in joint width. Splices in material are prepared by overlapping or butting joint ends. This material eliminates the need for joint backing, primers, and tooling.

Joint width determines the size of foam material required. If a joint varies considerably, more than 25 percent in width, different sizes of preformed foam sealant are required in the one joint.

A horizontal grade is available, allowing use in horizontal plaza and deck joints. Some have properties sufficient to withstand vehicular traffic as well. Foam sealant adheres to most clean and prepared building materials, including stone, aluminum, concrete, wood, and glass.

TABLE 4.9 **Precompressed Foam Sealant Properties**

Advantages	Disadvantages
No priming or backing	No colors available
Factory manufactured	Heat and cold affect installation
Nonlabor intensive	Difficult installation for varying joint width

In addition, foam sealant is compatible with other sealant materials and allows elastomeric sealants to be applied over the foam, providing a double barrier in critical waterproofing joints. With these applications, foam sealants are acting as a backing material for elastomeric sealants.

Most foam sealants withstand up to 25 percent movement in either direction for a total joint movement capability of 50 percent. Foam sealant performs best in compression mode with no long-term compression set, returning to its originally installed size.

Critical to successful foam sealant applications are well-cleaned joints. If the joint is wet or contaminated, the contact adhesive will fail. Materials are usually supplied in black only, but they can be painted to achieve other coloring, although paints will crack during movement. (See Table 4.9.)

Substrates

Successful sealant installation depends upon ensuring that a substrate is compatible with the material and is in acceptable condition for proper sealant adhesion. Adhesion is essential, without which all other properties are insignificant.

For example, Teflon®-coated materials, Kynar® finishes, or PVC substrates are especially difficult to adhere to. Teflon® is manufactured so that other materials do not adhere to it, thus keeping the surface continually clean. In these cases, butyl rubber sealant might be chosen over a silicone for its better adhesion capability, providing substrate movement is within the butyl's capability.

Sealant incompatibility with a substrate causes staining or etching of substrates. On the other hand, some substrates, such as oils, asphalt, and coal tar materials, may cause staining or sealant deterioration. To prevent these problems or when compatibility is in question, actual substrate testing with sealants should be done. This can be completed by testing under laboratory procedures such as accelerated weathering or by preparing mock-up panels with the sealant applied at job sites and allowing sufficient time to determine success or failure.

Following are descriptions of substrates commonly found in construction and their requirements for proper sealant installation.

Aluminum substrates

A common building component, aluminum substrates present difficulties when choosing sealants. This is due to the architectural coatings now being applied to aluminum to prevent aluminum oxidizing and to provide color.

By themselves, aluminum surfaces must be cleaned chemically or mechanically to remove any trace of oxidation that will prevent sealant adhesion. With coated aluminum, it is more difficult to choose a sealant.

Baked-on finishes and other coatings contain oils, carbon, and graphite residues, which act as release agents for sealants. Some coatings themselves may have poor adhesion to aluminum, thereby making it impossible to achieve proper adhesion with any sealant application. Other coatings may soften or deteriorate when solvents in sealants or primers come into contact with the finish.

The only positive method to test adhesion with a coated aluminum is actual testing before application. Silicones and butyls have acceptable adhesion to aluminum and are often used in aluminum curtain wall construction. But even these materials should be tested for proper adhesion. Elastomeric sealants such as urethanes are often used around aluminum frame perimeters and as such should be checked for adhesion, especially when coated aluminum products are used.

Cement asbestos panels

Cement asbestos panels are produced with a variety of finishes, including exposed aggregate and tile. The panels are attached to a wood or metal stud frame for support and attachment to a structure.

Often panels without finishes are less than ½ in thick. This composition prevents proper width to depth ratios for backing and sealant installation. Cement asbestos panels present difficult problems for sealant applications because of their high moisture absorption properties and thinness of the panel itself. After sealant installation any water absorbed into a panel can bypass sealant joints and cause damage to interior areas and panel support systems.

Cement asbestos panels have thermal coefficients similar to concrete panels, with movement at joints often exceeding movement capability of sealants. Compounded with absorption rates of panels, long-term performance of any sealant is questionable. In addition, form-release agents or sealers used on panels often contaminate joints, prohibiting proper adhesion of sealants. Some panels are manufactured with aggregate exposed in joint sides, which also prevents proper sealing.

These factors all contribute to problems in sealing and keeping panels watertight. To prevent such problems, adequate connections must be incorporated into panel design. Accelerated weathering testing of a panel design with wind and structural loading should be completed to verify the effectiveness of proposed sealant systems. Panel design should include details that make joints acceptable for sealant installation. These include joint sides at least 1 in thick, panel edges clean of any form-release agents or sealers, and aggregate not exposed on joint sides.

Precast concrete panels

Precast panels, including tilt-up and prestressed ones, are now produced in a variety of sizes, textures, and finishes. These have become a common building facing material for all types of structures. Problems arise not with the panels themselves but with sealers, finishes, or coatings applied to them.

Form-release agents are used in all precast panel fabrications. Since panel edges typically become sides of joints after erection, problems with adhesion occur. Oil- and petroleum-based products used for curing the panels will cause deterioration of silicone and polysulfide sealants. Film-forming curing and release agents can act as bond breakers between sealant and concrete.

Substrate adhesion testing often tests a sealant's ability to adhere to the form-release or curing agents rather than to a panel itself. Therefore, all precast joints should first be abraded or chemically cleaned to remove all residue of these compounds before sealant application.

Often precast panels are designed with exposed aggregate finishes. Although aesthetically pleasing, exposed aggregate often prevents proper joint sealing.

When panels are manufactured with aggregate turned or exposed onto panel sides (that later become joint sides), proper sealing is impossible. Sealants will not adhere properly to exposed aggregate, and the aggregate will prohibit proper movement characteristics of the sealant.

Figure 4.5 shows a typical improperly manufactured panel. If someone attempts to chip the stone out at a project site, pockets are created that must be patched with a cementitious grout. If such a repair is attempted, grout repairs can actually be pulled away from precast substrates during joint movement and cycling. The only acceptable repair method is to replace panels with precast panels, which are manufactured with no aggregate exposed within joints.

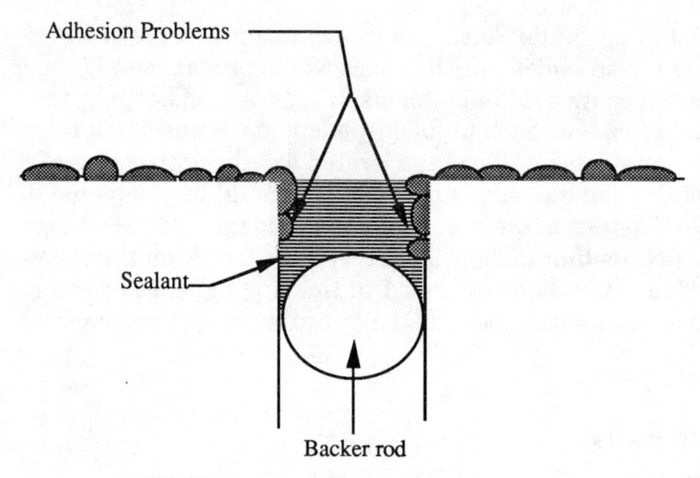

Figure 4.5 Exposed aggregate precast joint problems.

Project specifications often require coating application to panels after installation. These coatings include water repellents, antigraffiti coatings, color stains, and elastomeric coatings. Coatings applied before sealants can create problems. Sealant application should be completed first and protected during coating application; otherwise, sealants would be bonded to coatings rather than to the panel surface.

If a panel finish is porous and water absorption rates are high, water may enter the panel substrate and bypass sealant joints. Water causes bonding problems if faces of joints remain wet or if open-cell backing materials become saturated after sealant installation. This wet backing causes sealants to remulsify.

Wet substrates also cause the release of primers used for sealant installations. Absorptive panels require sealing with a water repellent after sealant installation to prevent these problems.

Panels should be erected and securely attached to prevent slippage, bowing, or creeping, which causes shearing and ripping of sealants. Panels must be installed so joints are uniform from top to bottom to prevent joints that are too narrow or too wide for proper sealing. In these situations an applicator may not bother to change the backing material size when the joint width changes, causing performance problems with sealants after installation.

Joints should also be kept uniform from one to the next. For instance, panels meant to have ½-in joints should not be installed with one joint 1 in wide and an adjacent joint virtually closed. Such variances will considerably shorten the life cycling of sealants.

Tiles

Quarry tile is manufactured with a patina finish, a result of firing tiles for smooth finishes. This finish should be removed by grinding joint sides before sealant application. Sealants should never be applied to grout in place of tile itself. Grout will eventually loosen and cause failures.

If efflorescence has formed on the tile before the sealing of joints, it should be removed chemically before applying sealants. Most manufacturers recommend primers when sealing quarry tile joints.

PVC

Polyvinyl chloride material such as PVC piping does not provide an acceptable substrate for sealant applications. It is necessary to mechanically abrade surfaces of PVC to be sealed. This roughens their surface before sealant application. This rough surface may provide an acceptable substrate for sealants such as butyl or silicones, but PVC materials should never be used at high-movement joint areas.

Stonework

Building facades of limestone, marble, or granite generally provide a surface acceptable for sealants. However, adhesion tests should be completed to determine their acceptability since there are so many finishes of each natural stone type available. Priming is usually required with these types of substrates.

It is important to note that in most cases a primer or uncured sealant may stain stone work. Therefore, precautions including masking joint faces before sealant or primer application will prevent staining. With porous absorptive stone, closed-cell backing material should be used to prevent backing from absorbing water that passes through the stone facade.

Terra cotta

Terra cotta tiles or stones manufactured from natural clay are typically supplied with a baked or glazed surface finish. However, sides of the tile are typically unfinished clay and are very porous and absorptive. Primers are required for adequate sealant bonding.

Should the facing of terra cotta be porous, water absorption may cause adhesion problems or gassing of sealants. Closed-cell backing materials should be used to prevent the backing from absorbing water entering through the terra cotta facade.

Sealant Application

Of all factors affecting sealant performance, installation is the most critical and most often causes joint failures. No matter how good a sealant is selected and how well a joint is designed, improper installation will lead to failures. Successful installation depends on several steps, including

- Joint preparation
- Priming
- Installation of backer rod or backing tape
- Mixing, applying, and tooling of sealant

Of these steps, the most common problem and most widely abused installation step is joint preparation. All remaining installation steps depend on how well this first step is completed. If joints are not properly prepared, regardless of how well joints are primed and sealed, materials will fail.

Joint preparation

The most common joint preparation problem arise when joints are not cleaned or when contaminated and incorrect solvents are used. All joint contaminants must be removed and joints must be dried before sealant and primer application.

To clean a joint, two rags are necessary—one rag to wet a joint with solvent, the other to wipe contaminants from the joint, while at the same time drying it. Using a single solvent rag will smear the contaminants in a joint. Continually dipping the same rag in a solvent will contaminate the entire container of solvent.

All loose mortar and aggregates must be removed since sealant will only pull loose material away from substrates when the joint moves. Other contaminants, such as waterproofing sealers, form-release agents, oils, waxes, and curing agents, must be removed. This may require mechanical methods such as grinding or sandblasting. It is important to note that after mechanical cleaning, joints must be recleaned to remove dust and residue left behind by mechanical cleaning.

Successful joint preparation steps include

1. Two-rag method of cleaning:
   Use lint-free rags.
   First rag has solvent poured on it, not dipped in solvent.
   Second rag removes solvent and contaminants and solvent.
   Change rags often.

2. Form-release agents, oils, paints, and old waterproofing materials must be removed by mechanical means, followed by recleaning joints, including pressure washing if necessary.
3. Joint sides must be dry and free of moisture or frost.
4. Loose joint sides must be chipped away and cut smooth. Jagged edges may cause air pockets to develop during sealant installation.

Priming

Primers are used to ensure adhesion between sealants and substrates. If there is any doubt if a primer is required or not, adhesive tests should be completed with and without primers to determine the most successful application methods.

Using too much primer, allowing primer to cure too long before installation, or applying sealants over wet primer will cause sealant failures. Manufacturer's application instructions pertaining to mixing, coverage rates, drying time, and application time vary with different types of sealant and primers. Instructions must be consulted on an individual basis for proper installation, including

- Use of proper primer
- No overapplication of primer
- Priming within application time recommendations
- Discarding of primers that are contaminated
- Manufacturer's recommendations

Backing materials

Backer rod and backing tape prevent three-sided adhesion in joint design. Tapes are used where a firm substrate against which to seal exists at backs of joints when joints have insufficient depth for backer rod installation. Rod is installed in joints where there is no backing substrate. The backer rod or backing tape provides a surface against which to tool material and maintain proper depth ratios.

Failure to install backer rod properly causes cohesive failure due to improper sealant width to depth ratio. Backer rod depth must be kept constant, which requires use of a packing tool. This simple tool, a roller that can be adjusted to various depths, is unfortunately rarely used.

Applying bond breaker tape incorrectly can cause joint failure both adhesively and cohesively. Tape allowed to turn up sides of joints will not allow sealant to adhere properly to sides, causing adhesion problems. If tape does not completely cover the backs of joints, three-sided adhesion occurs, causing cohesion failure.

Open-cell materials are usually made of polyurethane; closed-cell materials are manufactured from polyethylene. Sealant manufacturers will recommend the appropriate backing to use. Open-cell materials are not recommended for horizontal or submerged joints where water can collect in open cells. Closed-cell materials are inappropriate for moisture-cured sealants since they prohibit air from reaching the back of joint material.

Proper backer rod and tape installation depends on the following:

- Use of an adjustable packing tool, to ensure proper depth
- A backer rod that is 25 percent larger than joint width
- Backing material that corresponds in width to the varying widths of joints
- Horizontal joints without gaps in the rod that would allow sealant to flow through
- Bond breaker tape that covers the entire back of a joint but is not turned up at the sides
- Installation of tape over existing sealant to prevent three-sided adhesion (in remedial applications, where it is necessary to install new sealant over old)

Mixing, applying, and tooling sealants

Improperly mixed sealants will never completely cure and therefore will never provide the physical properties required. Improperly mixed sealants are evident from their sticky surface or softness of material, which can literally be scooped from a joint. All components of a material must be mixed in provided quantities. Never should less than the manufacturer's prepackaged amounts be used. Table 4.10 summarizes material coverage per gallon of material.

The use of proper mixing paddles and mixing for adequate lengths of time are important. Materials that have passed their shelf life, which is printed or coded on material containers, should never be used.

Proper application tools, including sealant application guns that are available in bulk, cartridge, or air-powered types, are necessary. The cartridge is used for one-component materials supplied in tube form. Bulk guns use two-component sealants. These guns should be filled carefully with the mixed material; air should not be allowed to mix with sealant, or gassing of materials will occur.

Nozzle selection and use is also important. Use a metal nozzle with a 45° angle and cut the plastic nozzles of tube containers to the same 45° angle. Joints might be overfilled or underfilled if proper nozzles are not used, causing tooling problems, improper depth to width ratios, and

TABLE 4.10 Approximate Coverage Rates for Sealant Materials

Depth (in)	Width (in)							
	1/16	1/8	1/4	3/8	1/2	5/8	3/4	1
1/16	4828	2464	1232	821	616	493	411	307
1/8		1232	616	411	307	246	205	154
1/4			307	205	154	123	103	77
3/8				137	103	82	68	51
1/2					77	62	51	39
5/8						49	41	31
3/4	Lineal feet of joint coverage per gallon of material						34	26
1								19

adhesions problems. Figure 4.6 shows substrate cracks being properly sealed before the waterproofing application.

Joints must be tooled to eliminate voids or bubbles and to ensure that the materials press completely against the sides of joints. Joints are tooled in a concave finish as shown in Fig. 4.7. This hourglass structure allows material to move properly and enhances the physical properties of a sealant.

Many types and sizes of tools are available for joint finishing, including those required for recessed joints. Soaps or solvents should never be used in tooling a joint because they will cause improper curing, adhesion failure, or color change.

Proper mixing, application, and tooling of sealants include

- Applying only in recommended temperature ranges, typically 50–80°F
- Mixing only complete packages of materials
- Using the appropriate mixing equipment
- Mixing for the proper amount of time
- Keeping air out of sealant during mixing
- Using properly sized nozzles and slopes to fill joints
- Tooling joints by compression for adequate adhesion
- Avoiding use of soaps or solvents in finishing joints

Cold weather sealing

Of the many problems that might occur in sealing joints in temperatures below freezing, the most serious is joint contamination by ice. In

Figure 4.6 Sealing of substrate crack for preparation of waterproofing application. (*Courtesy of Western Group*)

freezing temperatures, a joint surface can be covered with ice that is not visibly noticeable but that will cause the sealant not to bond to the substrate. Even if the sealant is warmed sufficiently to melt this ice, the resulting joint wetness will cause failure. Therefore, in freezing temperatures, it is critical that joints be heated and dried before sealant application.

Sealants in cold weather conditions should be stored in heated containers until the actual application. Curing time is slowed considerably, and sealants should be protected from physical abuse during this curing period.

With cold weather joint applications, joints are installed at their maximum width. These joints will always be in compression mode during movement and must be designed not to exceed the maximum width limit.

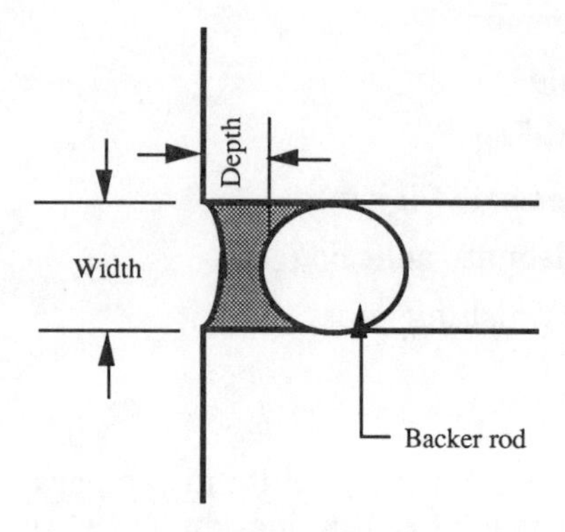

Figure 4.7 Typical sealant joint detailing.

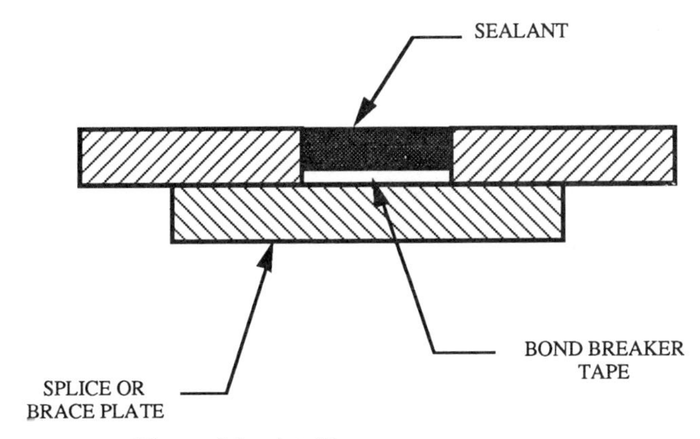

Figure 4.8 Narrow-joint detailing.

Narrow joints

Sealing thin or narrow joints, such as metal panels of curtain wall construction, presents several problems. The substrate area for sealant bonding is usually minimal, if not totally insufficient. Three-sided adhesion may be necessary if no allowance is available for application of a bond breaker tape.

For proper performance under these circumstances, a splice or backing plate of material should be installed behind the joint to allow for installation of bond breaker tape. In addition, sealants should be tooled flat and flush, not concave, which would leave a narrow section of material in the center. Refer to Fig. 4.8

Another alternative is to overband the sealant onto sides of metal facing as shown in Fig. 4.9. Note that in this situation, backing tape is brought up and onto the joint side to prevent three-sided adhesion. The

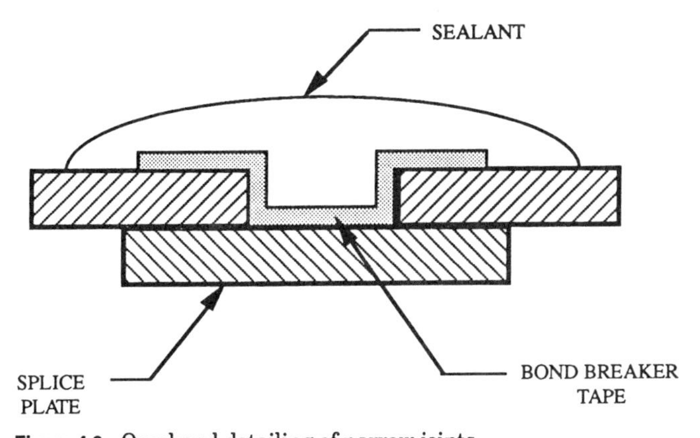

Figure 4.9 Overband detailing of narrow joints.

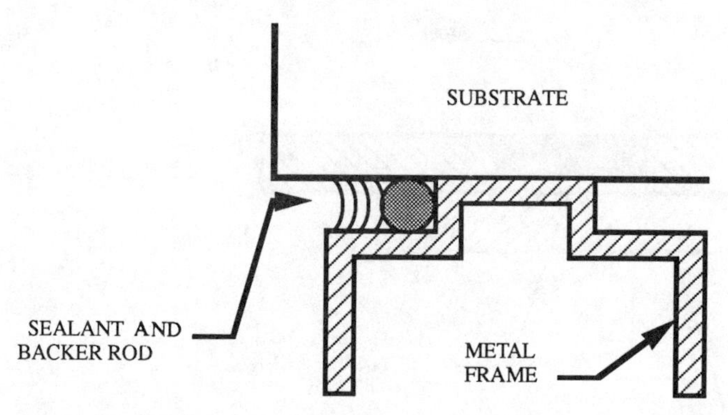

Figure 4.10 Metal-frame sealant detailing.

bonding area is determined by movement at a joint but should not be less than ¼ in.

Metal-frame perimeters

Sealing of metal-frame perimeters including doors, windows, and storefronts presents problems since rarely is a proper joint provided on which to apply the sealant. Typically, frames are butted up against surrounding structures including brick, precast, and curtain walls. If frames are smaller than openings, voids are left around the frame perimeters filled only with shims used for frame installation. In such instances, either there is no space to install backer rod or tape, or the frame is manufactured without sides against which to compress backer rod. This forces sealant installers to fill joints to incorrect depths, deterring joint effectiveness.

If a frame is butted to substrates, installers will usually place the sealant in a V-shape application by installing the sealant in a cant between the frame and substrate. This three-sided adhesion joint will not function properly. (See Fig. 4.11.)

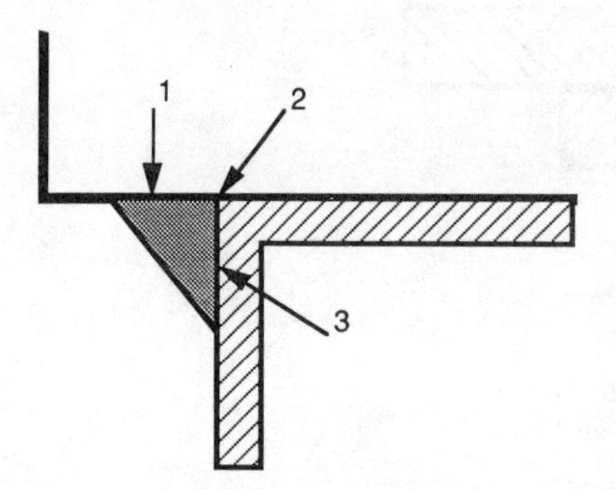

Figure 4.11 Incorrect metal-frame sealing.

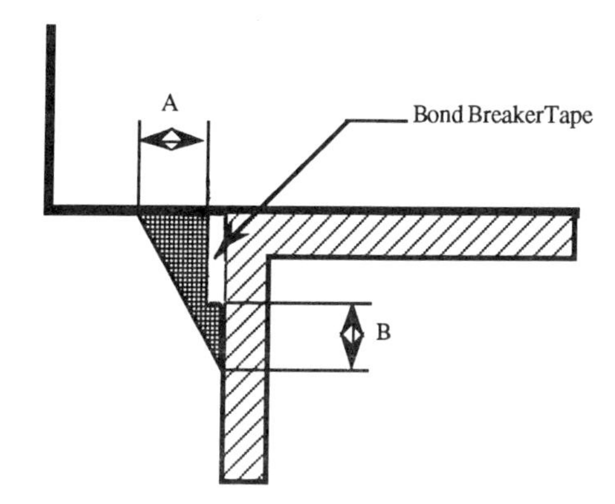

Figure 4.12 Correct metal-frame sealing. Distance A should equal distance B.

To seal these situations effectively, frames that allow a joint to be formed between frame and substrate should be manufactured, similar to that shown in Fig. 4.10. If this is not possible, for example, when repairing existing frame perimeters, steps must be taken to allow only two-sided adhesion. Sealant of equal thickness should be bonded to substrate and frame. (See Fig. 4.12.)

Bond breaker tape is installed as shown in Fig. 4.12 to allow contact area length to substrate equal to that of frames. Typically, the contact area length should be ½ in. In curtain wall construction or storefront perimeters, based on movement capability of sealant, this length may be increased to allow larger joints. Figure 4.13 shows glass perimeters being sealed.

Any gap or space between frame and substrate should first be filled with a caulking material. This provides a firm substrate on which to apply bond breaker tape. The sealant should then be tooled flat with a straight edge over the caulking. Table 4.11 summarizes the preferred installations for the major generic sealant materials.

Figure 4.13 Sealant application at glass perimeter. (*Courtesy of Western Group*)

TABLE 4.11 Generic Sealant Materials and Their Common Uses

Substrate	Acrylic	Butyl	Latex	Poly-sulfide	Poly-urethane	Silicone
Metal frame at interior	X	X	X			
Metal frame at exterior				X	X	X
Precast joints				X	X	X
Glazing and bed joints		X				X
Interior work	X	X	X			
Stucco crack repair	X		X			
Horizontal joints				X	X	
Submerged joints				X		
Wood joints			X			X
Metal curtain walls		X		X	X	X
Stone and masonry joints				X	X	X
Bath fixtures	X		X			X
High movement joints					X	X
Parking deck joints					X	
Marble					X	X
Granite					X	X
Limestone					X	X
Kynar® finish		X				

Chapter

5

Expansion Joints

The variety of expansion joints available is almost as numerous as their failures. Leakage is so common and failure so expected that expansion joints are available with integral gutters to channel the water leaking through joints. Manufacturers often recommend installing a gutter system below joints to collect leaking water. One only has to visit a few parking garages and view the numerous attempts at collecting leaking water to confirm this situation. Roof gutters, PVC piping, and metal collection pans are often used in makeshift fashion to collect water leakage.

Leaking water collects salts, efflorescence, lime, sulfites, and other contaminants as it travels through substrates. This contamination causes damage to automobile paint finishes and building structural components. There are numerous causes for expansion joints failure. Among the most prevalent are

- Selection of one joint for all details
- Improper detailing of joints into other building components
- Improper installation
- Use of too few joints
- Inadequate design
- Joints that are not capable of withstanding existing traffic

Expectation that one joint design will suffice for all conditions on a single project frequently causes failures. For example, a joint designed for horizontal straight runs is not appropriate for vertical installations, changes in plane, and terminations into walls or columns. Many joints are insufficient for 90° turns and changes in plane and often fail if such installations are attempted. Joint installations at walls or columns that abruptly stop with no provision for detailing joints into other building envelope components will fail. Attempts to install expansion joints continuously throughout a deck, including wall areas, planters, and seating areas, typically fail. Joints at building-to-deck intersections encounter considerable movement forces, including shear and differential movement, that often exceed joint capability.

For expansion joints to function properly over a range of in-place service requirements, they must include manufacturer details, design accessories, and systems components for the following common installations:

- Floor joints
- Wall-to-floor joints
- Building-to-floor joints
- Intersections with curbs
- Intersections with columns
- Joints at ramps
- Ramp-to-floor joints
- Intersection of two or more joints
- Changes in direction
- Joint terminations

Other common problems are connections between joints and substrates. These connections must withstand movement occurring at joints, or they will be ripped away from the substrate. If sufficient protection from traffic conditions is not provided, traffic wear over a joint might eventually break down or damage connections.

Often joints are not designed for the shear or lateral movement occurring in parking decks, especially at ramp areas. When an automobile travels over a joint, live loads induced by the automobile cause one side to lower while the opposite side remains level. Reverse action occurs after the auto passes over the joint. This *shear stress* can be felt by standing directly over a joint when automobiles cross.

Joints must be designed to withstand shear loading in addition to expected expansion and contraction movement. Expansion joints, such as T-joints, with a metal plate beneath the sealant portion, often fail because shear movement forces the plate into the sealant, ripping it apart.

For expansion joints to function when in place, they should also have the following components:

- Connection details for installation to structural components
- Connection details for waterproof coatings or membranes
- Protection against vehicular and snow plows
- Channeling of any water that might collect
- Cleaning provisions to remove accumulated dirt (e.g., leaves)

Expansion joints do not alleviate all movement encountered with deck construction. Concrete may crack short distances from expansion joints due to shrinkage, settlement, or differential movement. This is common with double-T precast construction incorporating a topping slab. At each panel joint, the topping slab is subjected to differential movement and will crack over each T-joint, regardless of how large an expansion joint is installed. Therefore, adequate allowances must be made for settlement and for differential and structural movement, which expansion joints alone cannot resolve.

For a joint to be successful it must have the following characteristics:

- The ability to withstand substantially more than the expected movement
- The ability to withstand all weathering conditions (e.g., freeze–thaw)
- The ability to withstand road salts and other atmospheric contaminants
- Facility of installation or training by manufacturer
- A superior connection to deck details
- Seamless along its length

In addition, adequate allowances for tolerances must be made in deck levelness.

Design of Joint

The first step in selecting an effective expansion joint is determining the amount of movement expected at a joint. This can be completed by computing the expected movement of a facade span or deck. This total movement is then divided into a number of strategically placed joints throughout the span.

Actual placement of the required joints is completed by a structural engineer. He or she determines where structural components can be broken to allow for movement in addition to where this movement is likely to occur.

Besides allowances for substrate movement, it must be determined what movement will occur in such areas as deck-to-building and floor-to-wall intersections. Differential movement and structural movement will occur at these areas, and an expansion joint system that will function under these conditions must be chosen.

For expansion joints at building-to-deck intersections, expansion material should be connected to both building and deck structural components, rather than facade materials. Expansion joints applied to surface conditions become loose and disbonded during weathering and wear cycles. Additionally, structural component movement may exceed movement capabilities of the facade causing joint failures. If it is necessary to install a surface-mounted joint, a secondary or backup seal should be installed beneath the expansion joint for additional protection.

In considering placement of joints, all design factors should be reviewed to avoid possible problems. For instance, it is not practical to place a planter, which is filled with soil and plants and is constantly watered, over an expansion joint. Even with proper protection, failure will occur when dirt contaminates the joint and disrupts movement capability. Furthermore, planter walls placed over a joint may not allow joint movement, causing failure of the joint and wall. Similarly, other items, such as equipment placement, column placements, light stanchions, and auto bumpers, should be reviewed.

Although movement amounts expected at joints is calculable, it is difficult to predict all types of movement that will occur. Factors such as wind loading, structural settlement, and distortion of materials impose directional movements that joints are not capable of withstanding. Therefore, selected joints should be capable of taking movement in any direction, a full 180° out of plane in all directions, to prevent failure.

Once a joint is selected and sized and appropriate accessories are selected to cover various details, proper joint terminations are designed. Simply stopping a joint at a wall, column, or termination of a deck will usually cause leakage. Attempting to apply a sealant over terminations is not sufficient. The sealant will not withstand movement that is likely to occur, especially if shear and other forces are encountered.

Manufacturers should provide specific termination detailing for complete weathertightness and movement capability at terminations. Any joints that channel water must incorporate allowances into the joint design to collect and dispose of the water at terminations. Most waterproofing systems are not manufactured to span expansion joints and are not capable of withstanding the movement that occurs there. Therefore, specific details must be designed for successful juncture of expansion joints into other waterproofing and building envelope components.

Choosing a Joint System

In choosing joint systems, examine all possibilities and choose a system for each specific need. Although convenient, it is not practical to choose one joint design for all conditions. Accordingly, manufacturers will have several types of systems and designs within each generic type to fulfill given project requirements. This prevents dividing responsibility among several manufacturers. Likewise, manufacturers of other building envelope components should approve the use of selected joint systems to ensure compatibility and complete envelope weatherproofing.

Generically, several systems are manufactured for use as expansion joints, including

- Sealant systems
- T-joint systems
- Expanding foam
- Hydrophobic expansion seals
- Sheet systems
- Bellows systems
- Preformed rubber systems
- Combination rubber and metal systems

Sealants

Sealants are often used as expansion joint materials, but they are only successful for joints with minimal movement. Sealants are not recommended for joint widths greater than 1 in. Joints larger than 1 in should be backed by other material such as expanding foam sealants or be used as part of the T-joint system. In designing sealant expansion joints, manufacturers recommend joint widths of four times expected movement when the material is capable of 25 percent movement in one direction.

Chapter 4 discusses the various sealant materials and their properties and uses. For horizontal expansion joint applications, polyurethane sealants are commonly used. Urethanes are capable of withstanding both pedestrian and vehicular traffic. They are compatible with deck coatings, sealers, and protected membrane applications.

Exposed sealant joints will not be effective when subjected to harsh traffic such as snow plows and vandalism. In such instances, it is advisable to protect sealants from abuse by installing a metal plate or other protection over joints. If protection is installed, it is attached to only one side of a joint to allow for movement. Refer to Fig. 5.1, for typical sealant expansion joint installation.

Sealant systems can be installed in a contiguous application with no seams. Terminations and junctures to other building envelope components are easily detailed and installed. Sealant systems are used ex-

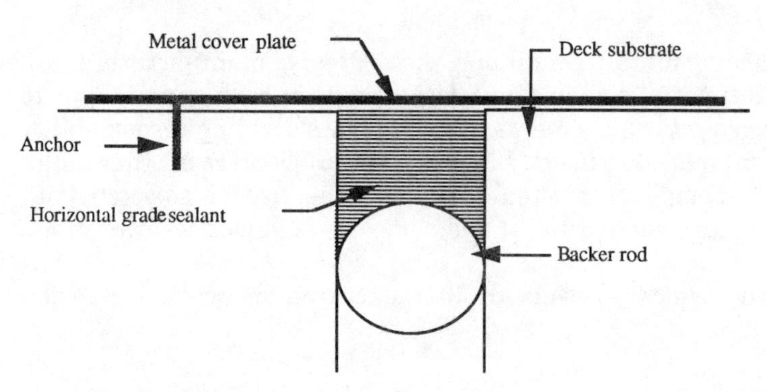

Figure 5.1 Sealant expansion-joint detailing.

tensively as expansion joint material with deck coatings and membrane waterproofing applications in which waterproofing systems are carried directly over the sealant. This installation type is effective as long as movement at joints does not exceed the capability of the waterproofing material. (See Table 5.1.)

For vertical applications, joint width also should not exceed 1 in unless specifically approved by the manufacturer. In joints exceeding 1 in, a backup material, such as expanding foam sealants, is suggested. In larger joints, it is advisable to cover sealants with a metal plate to protect against vandalism and excessive weathering.

Sealant materials are manufactured specifically for use in horizontal deck joints. They have properties making them resistant to traffic wear and contaminants such as oil, grease, and road salts. These materials are available for specific conditions such as airport runways. Typically, the material is a coal-tar derivative for resistance to gasoline and jet fuels.

In using sealant systems for remedial applications, all traces of previous material should be removed. If asphalt products were used, abrasive cleaning must remove all traces of contaminants. This will prevent problems associated with bonding of new sealants.

In all applications, joints must be protected during the curing stage, with no traffic allowed during this period, which may be as long as 72

TABLE 5.1 Sealant Expansion Joint Properties

Advantages	Disadvantages
Compatibility with waterproofing systems	Minimal movement capability
Seamless application	Protection recommended
Ease of terminations	Not for excessive wear areas

hours. Refer to Chap. 4 for specific application details for sealant joints.

T-joint systems

A T-joint system is a sealant system reinforced with metal or plastic plates and polymer concrete nosing on each side of the sealant. This system derives its name from a cross section of the joint, which is in the shape of a T. Figure 5.2 shows a typical T-joint configuration.

The basic T-joint consists of several components, including

- The sealant, usually urethane
- The reinforcement plate, ⅛-in aluminum or plastic reinforcement
- The bond breaker tape, which prevents three-sided adhesion to reinforcement
- The epoxy or polymeric nosing for attaching to substrates

Whereas sealant joints are only recommended up to 1 in widths, T-joint design allows for greater widths by adding reinforcement at both bottom and sides. With T-applications, joints as wide as 12 in (excluding nosing) are used. Design width of sealant in T-joints is recommended at five times anticipated movement, versus four times with regular sealant joints. This provides an additional safety factor for these size joints. Manufacturers require that joint width be not less than 3–4 in.

The T-system modifies a regular sealant joint to withstand the abuse and wear encountered in traffic-bearing horizontal joints. The metal plate reinforces the soft sealant during loading by automobiles. Nosings provide impact resistance and additional adhesion properties. It is recommended that this nosing extend approximately ⅛ in above the

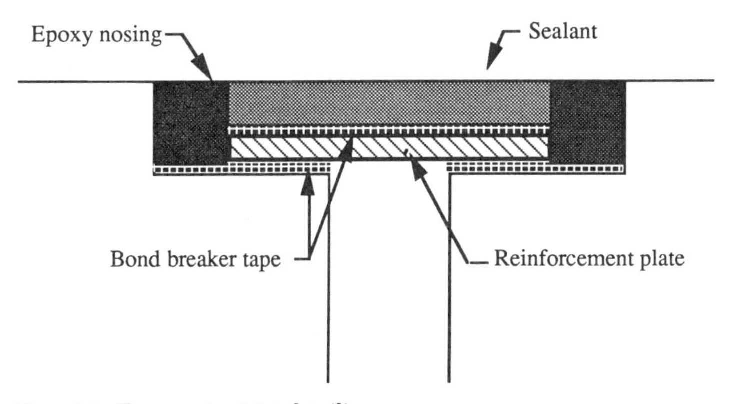

Figure 5.2 T-expansion-joint detailing.

sealant material and be sloped toward the sealant to prevent damage from automobiles and other heavy equipment.

The two installation methods for T-joints are fluid applied and preformed. Both use sealant materials, but a preformed system uses a sealant that has been formed and precured (see Fig. 5.3). This cured material is then placed into a joint at the job site. Fluid-applied systems are placed directly into a joint at sites after mixing and before curing (see Fig. 5.4). Both systems have distinct advantages and disadvantages.

Preformed systems allow for uniform sealant thickness and curing under controlled conditions. This prevents possible abuse that may occur during curing of fluid-applied systems. Preformed systems are not seamless applications and require a site filling of seams with compatible sealant.

Preformed material is usually formed in 8 ft lengths, requiring seams every 8 ft. Preformed systems do not provide allowances for irregularities with levelness of a substrate. The nosing is applied after the preformed sealant placement to alleviate any irregularities in joint width and levelness.

Fluid systems are vulnerable to damage and weathering during the curing stage. Colder temperatures may extend the length of typical

Figure 5.3 Sealing preformed T-joint seams. (*Courtesy of Tremco*)

Figure 5.4 Finishing of T-expansion joint. (*Courtesy of Tremco*)

curing time from 48 to 72 hours. Sealants used in expansion joints are typically a self-leveling grade. This causes fluid-applied sealants to flow to low ends of a joint, resulting in uneven joint thickness. Fluid systems may shrink somewhat in long joints and possibly pull away from the nosing during curing.

Bond breaker tape is required between a concrete deck and a reinforcement plate and between this plate and sealant. If bond breaker tape is installed improperly and turned up joint sides, improper adhesion will occur. Refer to Chap. 4 for further discussion of sealant tape installation.

Epoxy nosings are installed level with edges of the concrete deck and are installed by troweling to cover minor irregularities within the deck. With preformed joints, sealant is beveled along edges at approximately 45°, upon which nosing material is placed. This provides adequate bonding to secure the preformed sealant to a substrate. Nosing material should not flow or be troweled onto horizontal portions of the sealant surface, as this prevents sealant movement capabilities.

These joints may be applied over existing expansion joints by ramping the polymeric nosing upward to provide the required sealant depth. However, this exposes a joint to abuse by vehicular traffic and nosing will eventually wear, exposing sealant to damage.

TABLE 5.2 T-Expansion Joint Properties

Advantages	Disadvantages
Reinforced for better wearing	Labor intensive
Seamless application	Sealant portion exposed to wear
Ease of terminations	Not for excessive wear applications

T-systems are labor intensive, providing opportunities for job site misapplications as compared to factory manufactured systems that require minimal field labor. (See Table 5.2.)

Expanding foam sealant

Foam sealants should not be confused with generic sealants. Expanding foam sealants are composed of open-cell polyurethane foam, fully impregnated with a manufacturer's proprietary product formulation; these include neoprene rubbers, modified asphalts, and acrylic materials. Foam sealants are covered in detail in Chap. 4. A typical foam expansion joint is detailed in Fig. 5.5.

Foam materials are supplied in a compressed state, in rolls of various widths and lengths. For large widths, straight pieces 8–10 ft long are manufactured. A release paper over the adhesive on foam sealant facilitates installation.

These materials have considerably less elongation properties than better sealants, (150 versus >500 percent for sealants). They also have lower tensile strengths than sealants (20 versus 200 lb/in^2).

With limited elongation properties, these joints should be designed to be in a continuous compression rather than an elongation mode. Therefore, materials are provided in widths of two to five times the actual joint width, allowing materials to be in compression always.

Foam systems are particularly easy to install. The material is completely premanufactured and requires only that the joint be cleaned, contact paper removed, and the materials adhered to one side of the joint. Foam sealants then expand to fill a joint completely. Timing of this expansion is dependent on weather conditions, being slower in colder weather. These materials expand laterally and will not expand vertically out of a joint if properly installed.

Foam materials are extremely durable considering their low tensile strength. Once installed, foam is difficult to remove and is resistant to traffic and vandalism. Depending on the impregnating chemicals used, they can also be resistant to gasoline and oils.

Manufacturers produce several grades and compositions of materials designed for specific types of installations. These include below-grade

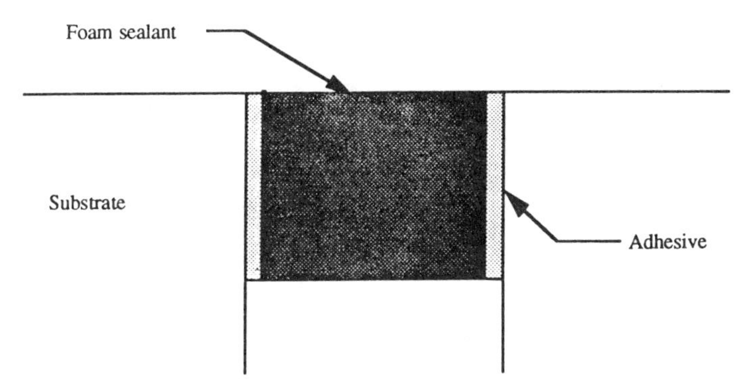

Figure 5.5 Foam expansion-joint detailing.

and above-grade joints, vertical or horizontal applications, and high-traffic grade for bridges and highways.

For vertical expansion joints, foam is often used as backup for a fluid-applied sealant. Horizontal installations do not require a cover plate or other protection. Foam sealants are also used as secondary protection in T-joints and are installed in place of standard backing material in a joint beneath the support plate.

Due to adhesion characteristics, foam material adheres to itself, providing seamless joint applications. It is recommended that joining ends of material be mitered for additional adhesion. These materials allow for 90° turns, with changes in plane, intersections, and terminations easily and effectively detailed. They are compatible with most building materials comprising the building envelope. (See Table 5.3.)

Hydrophobic expansion systems

Combining hydrophobic resins with synthetic rubber produces hydrophobic expansion seals. Hydrophobic refers to materials that swell in the presence of water. Thus, these materials require active water pressure to become effective water barriers. They are similar to below-grade clay waterproofing systems and therefore are limited to below-grade applications. As with foam sealants, materials are provided in

TABLE 5.3 Foam Expansion Joint Properties

Advantages	Disadvantages
Factory manufactured	Cost
Seamless application	Poor elongation
Ease of terminations	Low tensile strength

TABLE 5.4 Hydrophobic Expansion Joint Properties

Advantages	Disadvantages
Chemical resistant	Below-grade applications only
Follows contours of joint	Requires positive waterproofing systems
Good elongation	Used only with waterproofing systems

rolls in preexpanded form. Due to their reactivity with water, materials must not encounter water until after installation.

The use of hydrophobic expansion systems in expansion joints is extremely limited. Typically, they are used in conjunction with waterproofing membranes to fill expansion, control, or cold joints in below-grade construction. They are also used as waterstop materials in concrete substrates.

These materials swell from 2 to 10 times their initial volume. They have low tensile strength, but their elongation is similar to fluid-applied sealants, with some materials exceeding 500 percent elongation. As with foam, they should only be used in a compression mode. (See Table 5.4.)

Sheet systems

Sheet materials are manufactured from neoprene or hypalon rubber goods. They range from 40 to 60 mil in thickness and 4 to 12 in in width. Joint expansion and contraction are made weathertight by installing these materials in a bellows or loop fashion. This provides sufficient sheet material for stretching during contraction of a substrate. Material provided to form the bellows should be at least two to four times the expected joint movement. Figure 5.6 is representative of a typical sheet installation.

Materials are supplied in rolls 10–25 ft long. Seams are fused together by vulcanizing the rubber with a manufacturer's supplied solvent. Solvents are applied at seams that are lapped over each other, completely fusing the two pieces of material.

Materials may be applied to a substrate surface or recessed into a joint by installing a cutout along each side of the joint. Sheet systems are typically perforated along their edges for complete embedding of the sheet in an epoxy or polymer mix used to adhere the material to a substrate. This provides an effective mechanical bond allowing for installation over substrates other than concrete, including wood, metal, masonry, and glass.

Sheet installations allow materials to be applied at floor-to-wall joints, besides straight horizontal applications. Sheet systems are, however, difficult to install at transitions between a horizontal floor

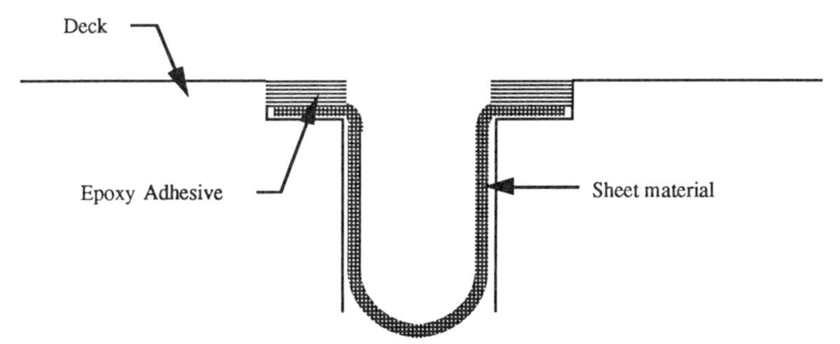

Figure 5.6 Sheet expansion-joint detailing.

joint and vertical wall joints. This is because of the bellows that forms in the material when making changes in direction or plane. It is not effective to turn the material in a 90° bend, as the bellows distorts and deters the system's effectiveness and bonding to a substrate.

Sheet systems are also difficult to terminate into structural components such as columns or ramp walls. At such details, material is formed into a box or dam and fused together. This design allows for collection of dirt and debris in the bellows that eventually prevents a joint from functioning. Further, water collecting in a bellows acts as a gutter with no drainage for water.

Because of these problems, joints should be covered with metal plates to prevent the accumulations. The cover plate also prevents possible safety hazards to pedestrians, who might trip on an exposed joint.

Sheet materials are effective choices in remedial applications. They can be surface mounted to an existing substrate, without requiring the substrate to be grooved or trenched for installation. Epoxy or polymer adhesives are applied in a ramp or slanting fashion to prevent blunt ends that might be damaged by vehicular traffic.

This type of installation allows the joint to be installed in applications where two different substrate materials, such as brick masonry, must be sealed. Refer to Fig. 5.7 for remedial detailing of a sheet system. Besides expansion and contraction movement, these systems also withstand shear and deflection (see Table 5.5).

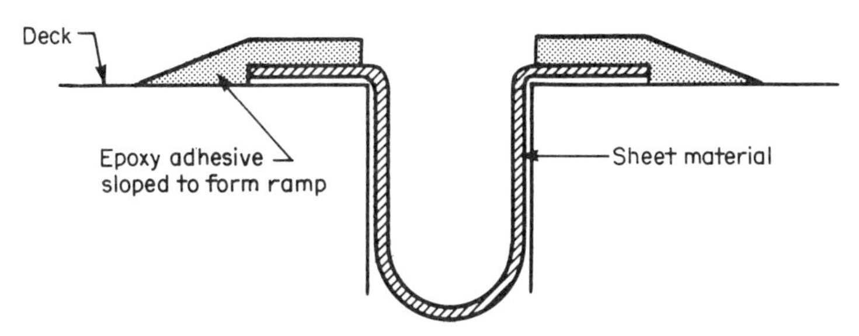

Figure 5.7 Remedial sheet expansion-joint detailing.

TABLE 5.5 Sheet Expansion Joint Properties

Advantages	Disadvantages
Vertical and horizontal applications	Collects debris in bellows
Good shear and deflection movement	Difficult to terminate
Metal, glass, and wood applications	Changes in-plane detailing

Bellows systems

Bellows systems are manufactured from vulcanized rubber into preformed joint sections. They are installed by pressurizing the joint cross section during adhesive curing that promotes complete bonding to joint sides. A typical installation is shown in Fig. 5.8. These systems are similar to preformed rubber systems but use air pressure for installation. Their cross sections are not stiffly reinforced by ribs manufactured in the material as are other preformed systems. Epoxy or polymeric nosing can also be installed to provide for better wearing at edges.

Bellows are available in sizes up to 3 in but are normally applied in joints 1 in wide. Joint material depth is approximately twice joint width. These systems function under 50 percent compression movement and 50 percent expansion movement.

Since bellows systems use preformed material, traffic can be applied immediately upon adhesive cure. Unlike sheet systems, adhesive is applied to interior sides of a joint, thereby protecting them from traffic wear. Additionally, the bellows is closed, which prevents accumulation of debris and water and therefore does not require a cover plate for protection.

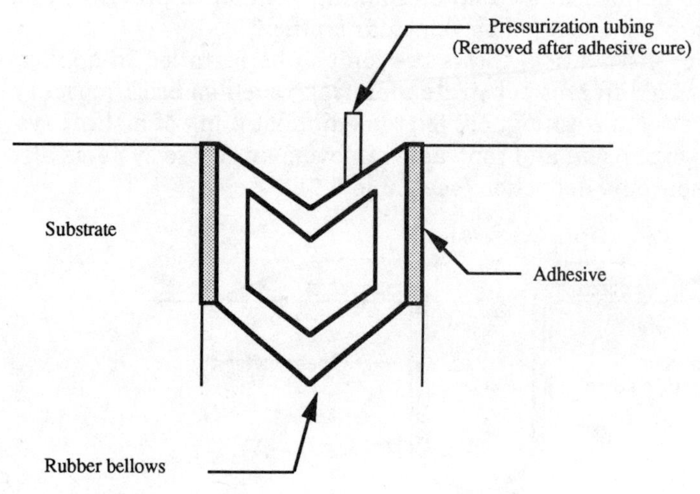

Figure 5.8 Bellow expansion-joint detailing.

TABLE 5.6 Bellow Expansion Joint Properties

Advantages	Disadvantages
No cover plates required	Maximum 3 in width
Factory manufactured	Difficult to terminate
No debris or water collection	Nonconforming to irregularities

These systems do not allow for major irregularities in joint width. This would prevent materials from performing in expansion or contraction modes. They also cannot take up irregularities in substrate unlevelness. This requires a joint be saw cut to uniform width and leveled before installation if necessary.

After adhesive is applied and bellows installed, air is injected to expand the joint cross section, similar to blowing up a balloon. This pressure is maintained until the adhesive is cured, at which time the pressure valves are removed and pressure holes sealed. This joint functions under movement in stress and deflection.

Material is supplied in roll lengths usually sufficient for seamless application. Should seaming be necessary, ends are vulcanized together with solvents.

Bellows systems are effective for surface-mounted floor-to-wall building joints. It is not, however, possible to turn the material 90° for changes in plane. Additionally, it is difficult to terminate these systems, and manufacturers should be consulted for recommendations.

For remedial installations, any existing joint material must be removed completely and interior joint sides must be cleaned for adhesive bonding. If existing joints are irregular in width or shape, they should be cut to uniform size for proper installations. (See Table 5.6.)

Preformed rubber systems

There are numerous preformed rubber systems available. These are manufactured from extruded synthetic rubbers such as neoprene and hypalon. They are available in countless cross sections and sizes. Unlike the bellows systems, they require a blockout or ledge in the substrate on which to place joint material. Many systems require an epoxy or polymeric concrete nosing at joint edges to prevent damage.

Many preformed systems have flanges attached to a compression seal that is perforated to allow for embedding into nosing material for mechanical bonding of joints to substrates. Other systems use metal frames for attachment to concrete substrates before concrete placement. Still others rely on chemical bonding to a substrate with adhesives.

Preformed rubber systems are available in widths ranging from ¼ to 6 in. Movement capability varies, but it is usually 50 percent compression and expansion movement. Rubber systems are very resistant to weathering and chemical attack from gasoline, oils, and grease.

These systems are typically used for straight horizontal runs only, but some are designed for use at vertical-to-horizontal junctures. These joints do not allow for 90° changes in plane. Some of the numerous cross sections of preformed expansion joints are shown in Fig. 5.9.

Preformed systems have high impact (tensile) strength, usually more than 1000 lb/in^2. This strength reduces movement capability, and materials should be sized accordingly. The high tensile strength allows for excellent wear resistance on areas subject to large amounts of vehicular traffic. Preformed systems do have limitations on the amount of shear and deflection movement they are able to withstand.

Only one portion of a preformed joint, the nosing, is job-site manufactured. This nosing anchors a joint to a substrate. Size of the nosing and adhesive contact area must be installed properly to ensure that the joint does not rip from the substrate during weathering or movement.

Flange Systems

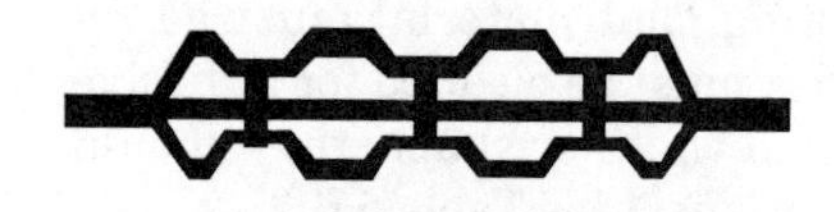

Adhesive Systems

Figure 5.9 Preformed rubber expansion joints.

TABLE 5.7 Preformed Rubber Expansion Joint Properties

Advantages	Disadvantages
Factory manufactured	Cost
High-impact strength	Difficult transitions
Chemical resistant	Limited remedial applications

Mechanical attachment of preformed systems is completed by using metal anchor bolts installed through holes in the rubber flange joint section. These systems require a blockout to allow a joint to be flush with a substrate. Anchoring should be checked by maintenance crews on a regular basis, since bolts may work themselves loose during joint movement. Horizontal seams are sealed by mitering ends of the rubber portion and fusing them with a solvent.

Premanufactured enclosures are available to terminate preformed joints into other building envelope systems. Factory manufactured accessories are available for changes in plane, direction, and intersections with other joints. Premolded joints are used in remedial applications but require concrete substrates to be cut to provide a ledge to place flanges. (See Table 5.7.)

Combination rubber and metal systems

Combination rubber and metal systems are manufactured with a basic rubber extrusion seal and metal flanges for casting directly into concrete placements. These systems are designed for new construction installations and are placed by the concrete finishers. They are not used in remedial applications unless major reworking of a deck is involved. Joints are manufactured in large sizes with no joints being less than 1 in wide.

The metal flanges on each side have reinforcement bars or studs for bonding with the concrete. Some combination joints include intermediate metal strands between the rubber for additional reinforcement and wear. Others include metal cover plates for additional protection against traffic wear.

Combination metal–rubber expansion materials may be interlocked to cover a joint width of more than 2 ft. These systems are costly and are used primarily in heavy service areas such as bridges and tunnels.

Combination systems require accessories for terminations, changes in plane or direction, and intersections with other joints. They must be carefully positioned before concrete placements and protected so as not to allow concrete to contaminate the joint. Concrete must be reinforced

TABLE 5.8 Combination Expansion Joint Properties

Advantages	Disadvantages
Available for large joints	New construction only
Factory manufactured	Cost
Extremely durable	For joints wider than 1 in

at the metal flange intersection to prevent substrate cracking and water infiltration that can bypass a joint.

In some instances, it may be advantageous to block out a joint location during concrete placement and install the joint with polymer concrete mix after concrete curing. A typical combination system is shown detailed in Fig. 5.10. (See Table 5.8.)

Vertical Expansion Joints

Most systems discussed thus far with the exception of sealant and foam materials, are manufactured for horizontal uses. In addition, there are several other systems available for vertical installations. These include premanufactured metal and plastic expansion and control joints, which are used primarily for stucco and plaster substrates.

Installations of stucco wall systems usually allow no dimension greater than 10 ft in any direction and no area larger than 100 ft^2 without control joints being installed for thermal movement. Joints also are installed where there are breaks in structural components behind stucco facades.

Preformed metal joints for stucco are available in a variety of designs and metals. The most durable metal is zinc, which does not corrode as materials such as galvanized metal do. Zinc materials withstand greater substrate movement than plastic or PVC materials.

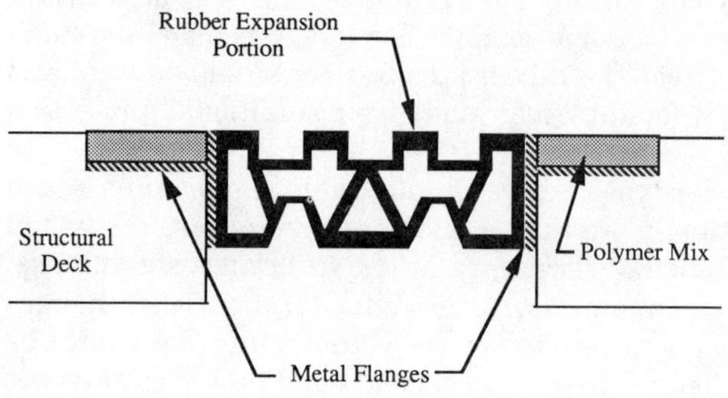

Figure 5.10 Combination rubber–metal expansion-joint detailing.

A typical cross section of a stucco control joint is shown in Fig. 5.11. The metal lath flanges are used to attach joints to substrates and are secured in place when stucco is applied over flanges. Flange sides should be secured to separate and structurally break the sides of a substrate to allow structural movement. Applying both flanges to the same structural portion will defeat the expansion joint purpose.

When vertical joints intersect horizontal joints in a facade, they should not be broken. Breaking the horizontal joint instead will prevent the water running down the facade from entering at joint intersections. These intersections should be monitored during installation as this is the most likely area of infiltration.

Stucco substrates often separate from preformed joints due to differential movement between the materials. This results in cracking along joint faces, allowing access for water infiltration into a structure and its components.

Unfortunately, to repair what might be perceived as a leaking joint, metal joints are often filled with sealant. This restricts joint movement capability and does not address the immediate problem.

In all types of vertical envelope surfaces, expansion and control joints should be placed at changes in plane or direction, at intersections of dissimilar materials, around substrate openings, and where allowances are made for thermal movement or structural movement.

Expansion Joint Application

All joint systems require that substrates be free of all dirt, oil, curing compounds, and other containments. Joints should be smooth, level, and straight to allow functioning and movement of expansion materials. With remedial applications, irregular areas should be sawn out, leveled, and chipped where required. If a ledge is necessary for installation, it must be free of all fins, sharp edges, and honeycombing.

Most joint systems, with the exception of those placed into concrete, require that the substrate be cured and dry. The various expansion systems require unique installation procedures as recommended by the manufacturer. Sealant and T-joints are applied as described in Chap. 4.

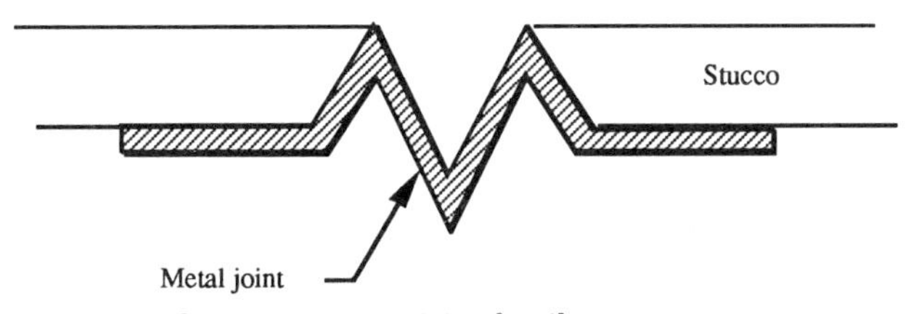

Figure 5.11 Stucco expansion joint detail.

Other expansion systems are factory manufactured and require only installation of adhesive or polymeric nosing.

Primers are generally required for horizontal sealant and T-joint applications. Polymer adhesives may require a solvent wipe or solvent primer before application.

Expansion joint systems have movement limitations. If deflection or shear movement is expected, use only materials expressly approved for this type of movement.

Chapter

6

Admixtures

Admixtures are used with masonry and concrete materials to enhance and improve in-place cementitious product performance. Admixtures are additions, other than normal ingredients of aggregate, sand, water, and cement, which impart desirable qualities to in-place concrete or masonry. These qualities might include

- Color
- Workability
- Shrinkage reduction
- Improved hydration
- Reduction of porosity
- Faster setting times
- Faster curing
- Waterproofing

Admixtures added during mixing of the concrete or masonry slurries add qualities throughout the in-place product. Surface-applied admixtures only disperse additional qualities to the substrate surface and depth to which it penetrates. Admixtures are available in many forms including

- Dry form
- Liquid additive
- Premixed cementitious form
- Dry shake or troweled-on (added at finishing stage)
- Liquid mixtures (applied during curing stages)

Hydration

Water added to cement, sand, and aggregate forms a paste that cures, hardens, and shrinks to create the finished concrete or masonry product. During curing, water leaves this paste through a process called hydration, which causes formation of microscopic voids and cracks in concrete. Once formed, these voids allow water absorption through the material.

Only controlled conditions of perfectly portioning, mixing, placing, and curing the concrete slurry will produce materials with minimum voids and absorption. Since field construction is never completed perfectly, however, concrete and masonry products often leak through the voids and cracks formed by the curing process.

The purpose of waterproofing admixtures is to provide complete hydration, which in turn promotes internal curing. This allows a reduction in shrinkage, providing a denser, higher strength and a more water-resistant product by reducing water absorption rates of a concrete or masonry material.

Admixtures available for concrete and masonry products that impart waterproofing or water-repelling characteristics include

- Dry shake
- Concrete admixtures
- Masonry admixtures
- Stucco admixtures
- Agents
- Polymer concrete

Dry Shake

The dry shake, power troweled, or shake-on methods use materials similar to cementitious membranes for below- and above-grade waterproofing. The difference is that unlike materials for cementitious membranes, dry shake admixture is applied during initial concrete finishing and curing (green concrete) rather than after curing. Shake-on admixtures consist of a cementitious base with proprietary chemicals that provide water-repellent properties.

These products are broadcast in powder form at ¾ to 1 lb/ft^2 of substrate area before initial concrete is set. Power troweling then activates proprietary chemicals with the moisture present in concrete.

With this method, the cementitious admixture becomes an integral part of a concrete substrate. These products do not merely add water repellency; they waterproof concrete against water-head pressure. They are effective admixtures used when waterproofing of concrete

TABLE 6.1 Properties of Dry Shake Admixtures

Advantages	Disadvantages
Simple installation	No movement capabilities
Above- and below-grade installations	Not completely waterproof
Becomes integral part of substrate	Must be applied during concrete finishing

substrates is required. These admixtures add compression strength to concrete substrates and abrasion resistance to withstand heavy traffic and wear. As with all cementitious systems, these products do not withstand cracking or movement in substrates by structural, thermal movement, or differential movement. (See Table 6.1.)

Dry Shake Application

Dry shake-on surface preparation requires only that concrete be in its initial setting stage, before power troweling. Areas being placed or finished should be no larger than a work crew can adequately cover by broadcasting material during this precured stage. Should concrete begin setting and curing, this method becomes ineffective for substrate waterproofing.

Use materials only in dry powder form as supplied by the manufacturer. Immediately after broadcasting, admixtures should be power troweled into the concrete. Dry shake products, as all cementitious products, are not used when moving cracks or joints are expected.

Masonry, Mortar, Plaster, and Stucco Admixtures

Masonry, mortar, plaster, and stucco admixtures are added directly to the water, cement, aggregate, and sand paste and are available in a liquid or dry powder form. They consist of organic chemicals, usually stearates, and proprietary chemicals, which impart integral water repellency. These admixtures lower the amount of water required for a paste mix, increase internal curing by increasing hydration, and reduce shrinkage. This results in a high density material, with high compressive strength, which absorbs substantially less water.

Specific additives, including chlorides, gypsum, metals, and other chemicals, might adversely affect concrete finishes or reinforcing in the substrate. For example, metallic additives bleed through finish substrates, causing staining, and increase chlorides, often leading to reinforcing steel deterioration. Therefore, product literature for each type of additive should be reviewed for specific installation procedures.

Admixtures typically reduce water absorption from 30 to 70 percent

TABLE 6.2 Properties of Masonry, Mortar, and Stucco Admixtures

Advantages	Disadvantages
Simple installation	No movement capabilities
Above- and below-grade installations	Can stain or damage substrate
Becomes integral part of substrate	Not completely waterproof

of regular mixes under laboratory conditions of controlled mixing and curing. Actual reductions, considering field construction inaccuracies, will substantially lower the results of water reduction.

Even with the high reductions of water absorption achieved, these products are not adequate for complete waterproofing of building envelope components. Also, their inability to resist cracking and movement further restricts their waterproofing characteristics. (See Table 6.2.)

Therefore, admixtures should only be considered as support or secondary measures in providing a watertight envelope. This includes admixtures added to mortar for laying brick veneer walls, which assists the primary waterproofing properties of the brick facade. Flashing, dampproofing, weeps, admixtures, and water-repellent sealers all become integral parts of the building envelope.

The admixtures' compatibility with primary waterproofing materials should be confirmed. Admixtures of this type may adversely affect bonding capabilities of waterproofing sealers or sealants. If in doubt, testing is recommended before actual installations are made.

Masonry and Stucco Admixture Application

Masonry and stucco admixtures require no specific surface preparation since they are added to the concrete, mortar, or stucco paste during mixing. Admixtures in quantities and mixing times recommended by the manufacturer should be monitored for complete dispersal throughout the paste. Water added to the paste must be measured properly so as not to dilute the admixture's capabilities and properties.

These materials are not waterproofing but water-repellent products. They will not function if cracking, settlement, or substrate movement occurs.

Capillary Agents

Hydration of concrete or masonry materials leaves behind microscopic pores, fissures, and cracks from water that is initially added to make the paste mixture. This hydration allows in-place concrete and ma-

TABLE 6.3 Properties of Capillary Admixtures

Advantages	Disadvantages
Simple installation	No movement capability
Above- and below-grade installations	Not completely waterproof
Fills minor fissures in substrate	Relies on chemical reaction

sonry materials to absorb moisture through these voids by capillary action. Capillary admixtures prevent this natural action and limit moisture absorption and water infiltration into a substrate.

Capillary admixtures are available in liquid or dry powder form that is mixed into the concrete paste, applied by the shake-on method, or rolled and sprayed in liquid form to finished concrete. Capillary admixtures react with the free lime and alkaline in a concrete or masonry substrate to form microscopic crystalline growth in the capillaries left by hydration.

A substrate should be totally damp to ensure complete penetration of capillary admixtures and provide the filling of all voids. This crystalline growth fills the capillaries, resulting in a substrate impervious to further capillary action. This chemical reaction requires moisture, either contained in a substrate or added if necessary.

As with other admixtures, these systems are not effective when cracks form in the substrate. Nor are they capable of withstanding thermal, structural, or differential movement. Capillary admixtures are further limited by the reliance of a chemical reaction necessary to form an impervious substrate. This reaction varies greatly depending on the following:

- Moisture present
- Alkali and lime available
- Admixture penetration depth
- Number and size of voids present
- Cracks and fissures present in a substrate

In the imperfect world of construction field practices, it is unrealistic to depend on so many variables to ensure the substrate watertightness that is essential to the building envelope. (See Table 6.3.)

Capillary Admixture Application

Dry shake or capillary admixtures are applied before the initial set and finishing of concrete. Admixtures added to concrete or mortar paste require no additional surface preparation. Liquid-applied materials require that concrete or masonry substrates be free and clean of all laitance, oil, curing, and form release agents.

Capillary admixtures chemically react with a substrate and require water for complete chemical reaction. Therefore, fully wet the substrate before application. At best, consider capillary admixtures as dampproofing materials, not complete waterproofing systems.

Polymer Concrete

Polymer concrete is a modified concrete mixture, formulated by adding natural and synthetic chemical compounds referred to as polymers. These polymers are provided separately to be added to a concrete paste or as a premixed dry form.

Although the proprietary chemical compounds (polymers) vary, the purpose of these admixtures is the same: They provide a denser, higher strength, lower shrinkage, more chemically and water-resistant concrete substrate. A comparison of typical concrete mixes versus polymer mixes is shown in Table 6.4.

Admixtures include chemicals to promote the bonding of polymer concrete to existing substrates. This allows polymer overlaying to existing concrete decks after proper surface preparation. These overlays are applicable as thin as ⅛ in thick, compared to at least 2 in thick for conventional concrete. This allows slopping of the polymer mix during installation to facilitate drainage and fill bird baths or water ponding on existing decks.

Additives also promote initial set and cure time, allowing substrates to withstand traffic in as little as 4 hours after placement. This can be very desirable in remedial or restoration work on parking decks.

Whereas capillary admixtures produce chemical reactions that fill the microscopic pores left by hydration, polymer admixtures produce reactions that eliminate or reduce these microscopic pores. Polymer

TABLE 6.4 Comparison of Regular and Polymer Concrete

Property	Regular mix	Polymer mix
Compressive strength	3000 lb/in^2	4000–8000 lb/in^2
Adhesive bonding	Poor	Excellent
Minimum thickness	2–4 in	⅛–½ in
Water absorption	High—10%	Low—0.1%
Chemical resistance	Poor	Good
Initial set time	72 hours	4 hours

TABLE 6.5 Properties of Polymer Concrete

Advantages	Disadvantages
Thin applications	Cost
High strength	Not completely waterproof
Chemical resistance	No movement capability

mixes also reduce the shrinkage that leads to cracking and fissures in a substrate allowing water penetration. These features provide the characteristics of low absorption that makes polymer concrete highly resistant to chloride attack. Polymer concrete products do not completely waterproof a structure. They are subject to cracking should structural, thermal, or differential movement occur.

Due to the high costs of polymer concrete, these materials are often used as overlays, not as complete substitutes for conventional concrete. These materials are used in renovations of existing concrete walks, bridges, and parking garage decks. They are also used for warehouse and manufacturing plant floors, where high impact strength and chemical resistance are necessary.

Polymer mixes are also chosen for installations not over occupied spaces, such as bridges, tunnels, and decks, where additional structural properties such as high compressive strength are necessary. In remedial installations such as parking decks, where reinforcing steel too close to the surface has caused concrete spalling, polymers provide an overlay to restore structural integrity. (See Table 6.5.)

Polymer Admixture Application

Polymer admixtures that are added to cement paste do not require any specific surface preparation. Polymer concrete applied as an overlay requires that existing substrates be thoroughly cleaned to remove all dirt, oil, grease, and other contaminants. Exposed reinforcing steel is sandblasted and coated with a primer or epoxy coating before overlay application. Existing decks should be thoroughly checked for delaminated areas by the chain-drag method and repaired before overlay application.

Polymer products added to concrete mixes require that proper mixing and preparations be used. Applications including overlays require proper proportioning and mixing according to the manufacturer's recommendations. Materials are placed and finished as conventional concrete. Working times with polymer concrete are substantially less than with conventional concrete.

In preparation for overlays to existing substrates, concrete should be sufficiently damp to prevent moisture from being absorbed from the polymer overlay mix. Polymers require no special primers. If a stiff mixture is required for application to sloped areas such as ramps or walls, a thin slurry coat of material is brush applied before the final application of the overlay (see Fig. 6.1).

These products are not applicable in freezing temperatures and are not designed as primary waterproofing materials of a building envelope.

Figure 6.1 Polymer concrete overlay application to existing parking deck. (*Courtesy of Western Group*)

Chapter

7

Remedial Waterproofing

Thus far the waterproofing systems discussed included applications for new construction or preventative waterproofing. Often, however, waterproofing applications are not completed until water has already infiltrated a building. Waterproofing applied to existing buildings or structures is referred to as remedial treatments or remedial waterproofing.

Leakage into structural components can damage structural portions and facades of a building envelope. In these cases actual repairs to a structure or its components is required before application of remedial materials. This type of repair is referred to as restoration. Restoration is the process of returning a building or its components to the original or near-original condition after wear or damage has occurred.

With historic restoration, new waterproofing materials or systems may not be allowable. Weathertightness then depends solely on a building's facade to resist nature's forces. Such facades are typically walls of stone or masonry. Unfortunately, this type of dependence may not completely protect a building from water damage, especially after repeated weathering cycles such as freeze–thaw cycles.

Many systems already discussed for preventative waterproofing may be used for remedial applications. In addition to these products, special materials are available that are intended entirely for restoration applications. If existing substrates are properly prepared, most products manufactured for new installations can also be used in remedial applications following the manufacturer's recommendations as necessary.

As with preventative waterproofing, in remedial situations no one product is available to solve all problems that arise. The availability of products used specifically for restoration is somewhat limited compared to that of the frequently used products of new construction applications. Applications and use requirements for preventative products are covered in Chaps. 2 through 6.

Remedial application needs are determined by some direct cause (e.g., leakage into interior areas). Restoration application needs are usually determined after leakage occurs or maintenance inspections reveal structural or building damage. In both cases, a detailed inspection report must determine the causes of leakage or damage and the repairs that must be made to a substrate or structure before waterproofing.

Once an inspection has been completed, causes determined, extent of damage reviewed, and systems or materials chosen, a complete and thorough cleaning of the structure or substrate is done. This cleaning may reveal additional problems inherent in a substrate.

Before waterproofing, application repairs to substrates must be completed since waterproofing materials should not be applied over unsound or damaged substrates. Once this preparatory work is complete, remedial systems should be applied by trained and experienced personnel (Fig. 7.1).

To reiterate, the sequence of events in remedial or restoration applications (as they differ from new applications), the following actions are necessary:

1. Inspection of damage and leakage
2. Determination of such cause
3. Choice of systems for repair
4. Substrate cleaning and preparation
5. Restoration work
6. Waterproofing system application

Figure 7.1 Surface preparation and repairs completed before application of remedial waterproofing treatment. (*Courtesy of Innovative Coatings*)

Inspection

Once an inspection is determined necessary, through either routine maintenance or direct leakage reports, a thorough analysis of a building's envelope should be completed. This analysis includes an inspection of all envelope components and their termination or connections to other components. This inspection determines causes of water infiltration and the extent of damage to building components (e.g., shelf angles).

Before the inspection, all available existing information should be assembled to assist in analyzing current problems. This information includes as-built drawings, specifications, shop drawings, maintenance schedules, and documentation of any previous treatments applied. Inspection of existing structures typically includes

- Visual inspection
- Nondestructive testing
- Destructive and laboratory testing

Visual inspection

Visual inspection may be done at a distance from envelope components (e.g., ground level), but preferably a close-up inspection is completed, which, if necessary, includes scaffolding a building. Scaffolding may also be necessary for actual testing of facade or structural components.

During visual analysis, documentation of all unusual or differing site conditions should be addressed. Visual inspections should locate potential problems, including

- Cracks or separations
- Unlevel or bulging areas
- Presence of different colors in substrate material
- Efflorescence
- Staining
- Spalled surfaces
- Missing elements

In addition to documentation of these areas, inspection should be completed on functioning areas of an envelope, including roof drains, scuppers and downspouts, flashings, and sealant joints.

Accessories available to complete visual inspections include cameras, video cameras, binoculars, magnifying glasses, handheld microscopes, plumbs, levels, and measuring tapes. The better the documentation, the better the information available for making appropriate decisions concerning repair procedures.

Either during visual analysis or after collection of data, further testing may be required to formulate repair procedures and document the extent of substrate and structure damage. Preferably, nondestructive testing, which does no harm to existing materials, will suffice. However, in some situations destructive testing is required to ensure that adequate restoration procedures are completed.

Nondestructive testing

Nondestructive testing is completed with no damage to existing substrates and typically requires no removal of any envelope components. Available testing ranges from simple methods, such as use of a knife, to advanced methods of x-ray and nuclear testing. The most prevalent nondestructive testing is water testing. In this analysis, water is applied by some means to a structure to determine areas of infiltration. Water testing is also used to measure moisture absorption rates of the various substrates that comprise a building envelope.

In conducting water tests, water is first applied at the base or bottom of areas being tested. Succeeding applications of water then begin upward. This prevents water from running down onto as yet untested areas. Water should be applied in sufficient quantities and time in one location to determine if an area is or is not contributing to leakage or absorption.

Once such a determination is made, testing moves to the next higher location. This testing requires someone to remain inside to determine when water leakage begins to occur. Water testing is limited in that it does not determine specific leakage causes or if leakage is created by damaged envelope systems within a structure such as cavity wall flashing.

Sounding is an effective means of determining areas of disbonding or spalling masonry materials. Such testing uses a rubber mallet to lightly tap substrates to discern differing sounds. For example, hollow sounds usually signify spalled or disbonded areas.

Another sounding method uses chains on horizontal concrete, masonry, or tile surfaces. This test is referred to as chain dragging. By pulling a short length of chain along a substrate, testers listen for changes in sounds, carefully documenting hollow sounds. The extent of those areas to be repaired is marked by painting or chalking an outline of their location.

Often, using a simple pocket knife to probe into substrates, without causing any permanent damage, can substantially supplement the information learned from visual inspection. A knife can be used to scrape along mortar joints to determine their condition. Should excessive mortar be removed, it is an indication of an underlying soft porous and poor strength mortar, which will require attention during remedial repairs. Knives can also be used for testing sealant joints by inserting the knife along joint sides to allow analysis of sealants and to determine if they are properly bonded to a substrate.

Water absorption testing is similar to water testing but only measured amounts of water are applied to a specific, premeasured substrate area over a specific length of time. This test can accurately determine absorption ratios of substrates. These results are compared to permeability ratios of similar substrates to determine if excessive absorption is occurring.

Modified laboratory testing at project sites can also be completed. This involves constructing a test chamber over an appropriate envelope portion. Static pressure testing as described in Chap. 10 can then be completed. Such testing requires an experienced firm that has the appropriate equipment to complete testing and the personnel to interpret test results.

Dynamic pressure testing of an envelope at project sites is also possible by using portable equipment that can introduce high air and water pressures. This allows conditions that simulate wind loading and severe rainfall to be applied against an envelope. Chapter 10 reviews jobsite testing and mock-up laboratory testing in detail.

Other testing devices include moisture meters, which give accurate moisture content of wood or masonry substrates, and x-ray equipment, which is used to locate and document metal reinforcement. Reinforcement can also be somewhat less accurately located by metal detectors and magnets.

More sophisticated equipment is available to determine existing moisture and its content in various substrates. This equipment includes infrared photographic equipment and nuclear moisture tests completed by trained and licensed professionals.

Destructive testing

Destructive testing involves actual coring or removal of substrate portions for testing and inspection. Examples include removing portions of a window wall to inspect flashings and surrounding structural damage and removing small mortar sections from a joint to test for compressive strength. Destructive testing is required when the extent of damage is not visually determinable or when deterioration causes are inconclusive from visual or nondestructive testing.

The most frequently used testing includes laboratory analysis of a removed envelope portion. Testing can consist of chemical analysis to determine if materials meet industry standards or project specifications. Testing can also determine tensile and compressive strength and extent of contamination by chemical or pollutant attack (e.g., by sulfites or chlorides).

Destructive testing includes probing of substrates by removing portions of building components to inspect damage to anchoring systems or structural components. Any removed envelope components should be reinstalled immediately upon completion of analysis to protect against further damage by exposing components to direct weathering.

Probing is also completed using a borescope. This equipment allows an operator to view conditions behind facade materials through a borehole only ½ in in diameter. The borescope comes equipped with its own light source allowing close-up inspection without removal of surrounding components or facing materials. (See Fig. 7.2.)

In-place testing is also used frequently, especially in stress analysis. Stress gages are installed at cracked or spalled areas, after which a wall portion adjacent to the gage is removed. Stress readings are taken before and after wall removal to determine amounts of strain or compression stress that was relieved in a wall after removal.

This test is helpful in such areas as building corners to determine if shelf angles are continuous around corners. These are areas in which stress buildup is likely to occur, resulting in settlement or stress cracking due to excessive loading.

Cause Determination and Methods of Repair

Analysis of compiled information from inspection results and any related data or documents are usually sufficient for a professional to determine leakage causes and the extent of damage. This analysis includes a review of all pertinent construction documents and maintenance records. Proper repair methods and materials can then be

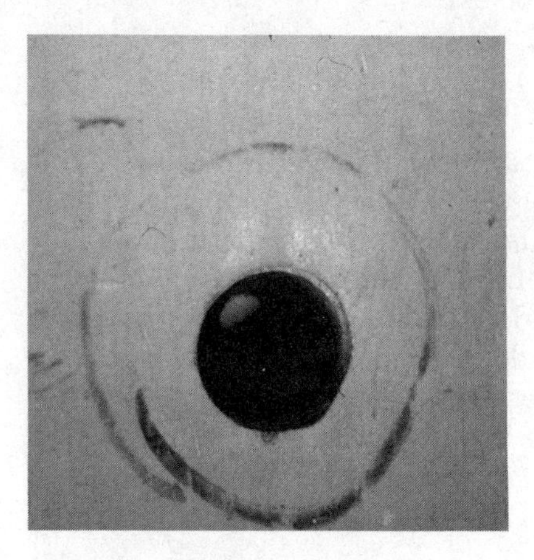

Figure 7.2 Destructive coring of masonry wall for investigation of structural elements.

chosen to complete remedial or restoration treatments. For example, if silicone sealants have previously been used in building joints, new sealants must be compatible with silicone or complete removal including joint grinding is required.

In reviewing test and inspection results, selecting repair procedures also depends on existing substrate conditions. For instance, if testing has revealed that mortar joints are allowing more water infiltration than existing dampproofing and flashing systems can adequately handle, a sealer application to masonry and mortar to prevent excess water infiltration may be required. However, if mortar joints are cracked or disbonded or have low strength, as determined by testing, a sealer application will not be successful. Additional repairs such as tuck-pointing would be required before sealer application to restore the envelope.

Determining water infiltration causes and choosing repair systems and materials should be completed by trained and experienced professionals. Prepared recommendations should be supplied to contractors for bid preparation. This ensures that all bids are prepared on the same basis of information, procedures, materials, and systems.

If recommended procedures are complicated or consist of several required methods to complete restoration, preparation of project specifications may be required. Specifications detail the types of products, materials, or systems to be used and the manner and location in which they are to be applied.

In addition to specifications, drawings may be required to show repair locations and their relationship to adjacent building envelope components. This enables contractors to prepare adequate bids for completion of remedial repairs and the restoration project. Any additional information that is useful to bidding contractors, such as as-built drawings, original job specifications, and access to site for reviewing existing job site conditions, should also be provided.

All completed repairs or restoration work should be carefully documented. This documentation should be maintained by the building owner for future reference should future repair or warranty work be required. Once these remedial procedures are determined and specifications are prepared, the next step in the restoration process is building cleaning.

Cleaning

Existing structures have surface accumulations of airborne pollutants that must first be removed to allow bonding of remedial waterproofing materials. If surface pollutants and dirt accumulations are considerable, it will be necessary to require building cleaning before inspections. This provides adequate conditions to review present conditions and make appropriate decisions.

Besides aesthetic purposes, cleaning is a necessary part of building maintenance. Maintenance cleaning ensures proper life cycling and protection of building envelope components against pollutant contamination.

Pollutants affect envelopes in two distinct manners. The first is by direct substrate deterioration by pollutants, including salts, sulfites, and carbons that, when mixed with water, form corrosive acids including sulfuric acid. These acids are carried into an envelope system in liquid state. Acids attack not only in-place waterproofing systems but also structural components, such as reinforcing steel or shelf angles.

The second pollutant deterioration is the slowing or halting of the natural breathing process that allows entrapped moisture to escape. Pollutants carried deep into substrates by water and moisture fill microscopic pores of envelope facades. If left unchecked, this collection will stop natural moisture escape that is necessary in substrates and will lead to damage from freeze–thaw cycles, disbonding of coatings, and structural component deterioration.

Building cleaning is therefore a necessary part of building preservation and proper maintenance. Cleaning should not be put off until remedial and restoration applications are necessary.

Exterior cleaning is completed by one or a combination of these methods:

- Water cleaning
- Abrasive cleaning
- Chemical cleaning
- Poultice cleaning

Before cleaning by any of these methods, testing of the proposed method is imperative. Testing ensures that cleaning systems are adequate for the degree necessary and that damage to the existing substrates, adjacent materials, and existing waterproofing systems will not occur. Sample testing a lower building portion in areas away from high traffic and visibility is desirable.

Cleaning is normally completed before starting remedial repairs. In some situations, however, a building can be in such a deteriorated state that introduction of chemicals or water under pressure will further damage interior areas and structural components. In these situations, sufficient remedial repairs may be required before cleaning to prevent further damage.

Water cleaning

Building cleaning done by water includes pressure washing, water soaking, and steam cleaning. Pressure washing is the most common procedure, especially when it is used in conjunction with preparation of substrates for waterproofing applications. It is also used in combination with other cleaning methods such as aggregate and chemical cleaning.

Pressure cleaners are manufactured to produce pressures varying from 300 to more than 25,000 lb/in^2. The lower pressure cleaners are used for rinsing minor residue accumulation, whereas the higher pressure machines remove not only pollutant collection but also paints and other coatings. The wide range of pressures available require that testing be completed to determine the pressures required to remove contaminants without damaging substrates.

Equipment spray tips and operators directly control pressure cleaning results. Fog-type spray nozzles are desirable and impart little harm to a substrate, whereas O-tip nozzles greatly concentrate the stream of water and can cause substrate damage.

Applicators must be experienced in this type work, especially with higher pressure equipment. Unskilled mechanics can damage a building by blowing sealants out of joints, damaging masonry or stone finishes, leaving streaks, and performing insufficient cleaning in the process. (See Fig. 7.3.)

Water soaking is a cleaning method preferred by preservations and historical restoration professionals due to the minimal amount of damage possible. Water soaking is especially successful on limestone structures, where chemical or pressure washing is unacceptable.

Specially prepared soaker hoses or sprayers are installed on upper building portions and provide a continuous curtain of water flowing down the building facade. After an initial period of soaking, determined by actual project testing, this method loosens dirt and pollutant accumulations. These pollutants are then removed by low-pressure spray cleaning. In highly contaminated areas, a repeat soaking process may be necessary to clean a building to acceptable aesthetic and project requirements.

A major disadvantage to water soaking is the amount of water introduced onto the exterior envelope. If deterioration or leakage is occurring, this system will cause further damage. Soaking will also deepen penetration of salts and other contamination into pores of a substrate, which follow-up pressure washing may not remove.

Available water supplies often contain minerals that stain or streak existing substrates. Water purification equipment is necessary to treat water before application.

Surfaces being prepared for waterproofing applications by using

Figure 7.3 Pressure cleaning operation. (*Courtesy of ProSoCo*)

water soaking require a long drying period. As long as 1 month may be necessary for substrates to dry sufficiently. Also, to prevent damage, preliminary remedial waterproofing, such as tuck-pointing, will be required before start of soaking.

Steam cleaning, although used extensively in the past, is now almost obsolete due to expanding technology in pressure cleaning equipment. Steam equipment rapidly heats water in a self-contained boiler; then it spray applies water under low pressure. The heated water swells and loosens collected pollutants, enabling them to be rinsed off a substrate.

Results achieved by steam cleaning are now reachable by pressure equipment that is much lower in cost than steam equipment. However, particular substrates can be so contaminated that too high a pressure may be required to achieve results obtained by steam cleaning. (See Table 7.1.)

Abrasive cleaning

Abrasive cleaning uses an abrasive material to remove mechanically accumulated dirt and pollutants (see Fig. 7.4). Abrasive cleaning methods include

- Sandblasting
- Wet aggregate blasting
- Sanding belts
- Wheel grinders

TABLE 7.1 Water Cleaning Properties

Advantages	Disadvantages
Several methods available, including pressure and soaking	Introduces water to envelope components
Chemicals can be added if required	Improper cleaning can damage substrate or cause streaking
Variable pressures	Environmentally safe

Abrasive cleaning systems remove not only surface accumulations of dirt but also some portion or layer of a substrate itself. This surface damage is often severe, and substrate restoration may be necessary. These systems are not preferred for substrate preparation, waterproofing applications, or general building cleaning and are not acceptable to most restoration and preservation professionals.

Abrasive cleaning is now typically limited to paint removal on metal substrates, although this procedure is now economically possible with advanced technology in water blasting equipment and chemical paint removers.

Wet aggregate cleaning is the mildest abrasive cleaning process. This method uses a vortex attachment on a pressure cleaner that suctions sand to mix with water at the spray tip. Water pressure then directs the aggregate against a substrate. This method operates under

Figure 7.4 Abrasive cleaning process. (*Courtesy of Western Group*)

TABLE 7.2 Abrasive Cleaning Properties

Advantages	Disadvantages
Removes paint layers easily	Can damage substrate excessively
Flour sand reduces damage	Safety concerns
Variable pressures	Equipment required

lower pressures than compressed air-blasting equipment and also wets the aggregate, keeping airborne contaminants to a minimum.

With all abrasive cleaning, some portion of a substrate surface will be removed. Careful testing should be completed to analyze the process before complete substrate cleaning. Additionally, because of potential damage and safety concerns, only highly experienced mechanics should be employed in these cleaning processes.

By using extremely small aggregates, substrate damage is lessened but still produces desired cleaning results. A very fine powdered sand, referred to as flour sand because it has the consistency of baking flour, is available. By using this sand with low pressures, satisfactory results with insignificant substrate damage are possible.

Sanding and mechanical wheel grinders are used to remove paints and corrosion from metal substrates. Grinders also have limited usage on concrete substrates for removing small contaminated areas of oil, grease, and other accumulations, which pressure cleaning will not remove. This cleaning also removes portions of the substrate and should be used only when other alternatives are not available. (See Table 7.2.)

Chemical cleaning

As with water pressure cleaning, chemical solutions are available in a wide range of strengths for cleaning. Substrates infected with specific stains not removable with plain water can be chemically cleaned. These cleaners include mild detergents for mildew removal to strong organic cleaners for paint removal (Figs. 7.5 and 7.6). Chemical cleaning formulations include three types, all of which include a manufacturer's proprietary cleanser:

- Acidic
- Organic
- Alkaline

Cleaners are toxic and should be used by trained personnel wearing protective clothing. Cleaners should be applied in sample areas so that damage such as etching of stonework does not occur. Adjacent envelope

Figure 7.5 Chemical cleaning process. (*Courtesy of ProSoCo*)

components, including glass, metals, and vegetation, should be completely protected before cleaning.

OSHA, EPA, state, and local regulations control chemical cleaner usage, including their collection and disposal. Most municipalities will not allow chemicals to reach city drainage, surrounding soil, or groundwater. Some cleaners have formulations that are neutralized after rinsing with water; others do not. It is important to investigate manufacturer's recommendations and local codes to prevent unlawful use or disposal of chemical cleaners. Refer to Chap. 9 for additional hazardous waste use and disposal regulations.

Chemical cleaners are necessary when water cleaning does not suffice and abrasive cleaners cause too much substrate damage. Removing paint with chemical cleaners only requires rinsing to remove paint residue after cleaner application. It is often necessary to repeat applications several times when previously painted layers are excessive or several different paints have been applied. With lead-based paints, waste from cleaning must be treated as hazardous waste, properly collected, removed, and disposed according to government regulations.

Cleaners are also available for stain and pollutant removal from substrates. These substances include asphalt, tar, and metallic and efflorescence stains. The cleaners remove specific areas of stains on a

Figure 7.6 Paint stripping using chemical cleaners. (*Courtesy of ProSoCo*)

substrate in conjunction with general pressure cleaning or soaking (Table 7.3).

Poultice cleaning

When existing stains or pollutants have penetrated a masonry surface, water and chemical cleaning are often not sufficient to remove staining. If abrasive cleaning is not acceptable, poulticing may be an alternative method. With poultice cleaning, an absorbent material such as talc, fuller's earth, or a manufacturer's proprietary product is applied to a substrate. This material acts to draw stains out by absorbing pollutants into itself. The poultice is then removed from the substrate by pressure cleaning.

The length of time a poultice must be left on a substrate to absorb

TABLE 7.3 Chemical Cleaning Properties

Advantages	Disadvantages
Little damage to substrate	Environmental and safety concerns
Ease of paint removal	Clean-up and disposal requirements
Various formulations and strengths available	Damage to surrounding substrates and vegetation

TABLE 7.4 Poultice Cleaning Properties

Advantages	Disadvantages
Removes deeply penetrated pollutants	Requires extensive technological knowledge
No damage to substrate	May force pollutants deeper
Excellent for natural stone substrate	Extensive testing required before application

pollutants varies with the stain type, pollutant penetration depth, substrate porosity, and general cleaner effectiveness. This cleaning is especially effective on natural stone substrates such as limestone, marble, and granite. Poultice-type cleaners are effective on a wide range of stains, including oil, tar, primer, solvents, paint, and metallic stains from hard water (Table 7.4).

Substrate testing with various types of available cleaning systems should be completed to determine the most effective system that does no harm to facade and adjacent materials. Complete and thorough cleaning of substrates is necessary before proceeding with the restoration phase.

Restoration Work

Just as remedial waterproofing systems must be applied over clean substrates, they must be applied over sound substrates. After cleaning, all restoration work must be completed before waterproofing material is applied. Any substrate deterioration that has occurred, including spalled concrete, damaged structural components, and oxidized reinforcement steel, should be prepared.

Restoration work often requires removal of building envelope portions to repair structural deterioration. This includes anchoring devices, pinning, and shelf angles used for attaching facing materials to structural building components.

This repair work is necessary after years of water infiltration high in chloride content, which corrodes metal components. Other required repairs, including control or expansion joints installation and cleaning of weep holes, are also completed at this time.

After completing all necessary repairs and substrate preparation, remedial waterproofing systems installation can begin (Fig. 7.7). Preventative waterproofing materials, discussed in Chaps. 2 and 3, can be applied as remedial treatments if the surfaces are properly prepared. Remedial treatments also include installation of flashing, sealants,

Figure 7.7 Cleaning completed before restoration work commencing. (*Courtesy of ProSoCo*)

and other envelope transitional materials found to be inadequate in the original construction.

In addition to preventative waterproofing systems, several waterproofing materials and systems are manufactured specifically for remedial and restoration projects. In some cases, even before a building is completed, remedial products are required to repair damage occurring during construction.

Remedial waterproofing systems now available include

- Tuck-pointing
- Face grouting
- Joint striking
- Mass grouting
- Grout injection
- Epoxy injection
- Cementitious patching
- Shotcrete and gunite

Tuck-pointing

In most masonry structures, unless the masonry was handmade and is excessively porous, any leakage is usually attributable to mortar joints. The water, moisture, or vapor that passes through the masonry itself is usually repelled by dampproofing or flashing or weep systems.

Through the aging process, all mortar joints eventually begin to deteriorate, caused by a multitude of weathering factors. These include swelling of masonry, which when wetted, places pressure on mortar joints from all sides. This causes fractures and cracks along the masonry and mortar junctures. Other factors contributing to mortar deterioration include freeze–thaw cycling, thermal movement, and chemical deterioration from sulfites and chlorides in atmospheric pollutants.

During life cycling, weathering begins to allow significant amounts of water and moisture through a masonry wall. Eventually this water may exceed the capabilities of existing dampproofing systems, allowing water to penetrate interior areas. Entering water also begins structural deterioration behind masonry facades.

If building maintenance inspections reveal that mortar deterioration is contributing to excess water infiltration, tuck-pointing mortar joints will be necessary. Tuck-pointing is a restoration treatment used to restore the structural integrity of mortar joints. Tuck-pointing procedures include removing existing deteriorated mortar and replacing it with new mortar (Table 7.5).

Inspections may reveal that only certain wall joints require tuck-pointing, or an entire wall area may require complete tuck-pointing to restore the building envelope. If miscellaneous tuck-pointing is required, specifications or bid documents should be explicit as to what constitutes sufficient deterioration to require removal and replacement. The tuck-pointing type of repair requires inspection to ensure that deteriorated joints are being repaired as per the contract.

For complete tuck-pointing projects all joints will be restored, but inspection procedures should also be structured to ensure that all joints are actually tuck-pointed. Economics of complete tuck-pointing often lead to considering alternate repair methods, including face grouting or complete regrouting.

TABLE 7.5 Tuck-Pointing Properties

Advantages	Disadvantages
No aesthetic changes to substrate	Labor intensive
Environmentally safe	Cost
Repairs can be limited to a specific area	Mortar removal may damage surrounding masonry

Tuck-Pointing Application

Masonry walls should be thoroughly checked for contaminants before tuck-pointing. Existing mortar should be removed to a minimum depth of ⅜ in, preferably ½ in. Up to 1 in removal of severely deteriorated joints is required. These depths allow bonding between existing and newly placed mortar and the masonry units.

Joint removal is completed by hand or with power tools such as hand grinders (see Fig. 7.8). On historic structures or soft masonry work, power tools damage existing masonry too extensively. Power tools often cause irregular joint lines, or actual portions of masonry might be removed. Sample areas on older masonry structures should be completed for analyzing acceptability of power tool usage.

Once defective mortar is removed, joint cavities must be cleaned to remove dust and mortar residue. This residue, if left, will deter the effective bonding of new mortar. A preferred method of residue removal is spraying joints with compressed air.

Once preparatory work is completed, existing mortar cavities should be wetted just before tuck-pointing application. This prevents premature drying and curing, which results in structurally weak joints.

Only premixed materials specifically manufactured for tuck-pointing should be used. These dry mixed cement and sand-based products contain proprietary additives for effective bonding and waterproofing and are nonshrinking. Materials higher in compressive strength than the masonry units are not recommended. If joints are stronger than the

Figure 7.8 Tuck-pointing application. (*Courtesy of Western Group*)

masonry, spalling of masonry units during movement in the wall system will occur.

Materials should be mixed using only clean water in amounts specified by the manufacturer. Pointing materials are available in premixed colors, or manufacturers will custom match existing mortar. Field mixing for color match should be prohibited, as this results in inadequate design strength and performance characteristics.

Pointing mortar must be applied using a convex jointer that compresses and compacts material tightly into joints and against sides of masonry units. This creates an effective waterproof mortar joint. The tooler or jointer should be slightly larger than joint width, and enough mortar should be placed in joints so that after jointing excess material is pushed from joints. This ensures that joints are properly filled to capacity.

After initial mortar set, joints should be brushed or scraped to remove fins formed by applying this material overage. Finished joint design should be concave or weathered for longevity and weathertightness. Refer to joint design in Chap. 8.

Priming of joints and bonding agents is not required. Dry mixes supplied by manufacturers contain all necessary components. Pointing should not be applied in conditions under 40°F or over extremely wet surfaces.

Face Grouting

Certain restoration projects include deteriorated masonry units requiring remedial procedures for both joints and masonry units. In a process referred to as face grouting or bag grouting, a cementitious waterproofing material is brushed and scrubbed into mortar joints and masonry faces. This grout is then brushed off just before complete curing of grout.

Grout materials are cement- and sand-based products with proprietary waterproofing chemicals and bonding agents. Some materials contain metallic additives that may change the color of a substrate when metallic materials begin oxidizing. Manufacturer's data should be reviewed to judge product suitability for a particular installation.

Bag grouting refers to a technique using burlap bags to remove grout after application to wall areas. Grout is used to fill pores, cracks, and fissures in both the joints and masonry, waterproofing an entire wall facade. Face grouting does not change the color or aesthetics of wall surfaces nor the breathability of facing materials. Face grouting will, however, impart a uniform color or shading to walls; the effects depend on the grout color chosen. Testing of sample areas should be completed

TABLE 7.6 Face Grouting Properties

Advantages	Disadvantages
Repairs both masonry and joints	Cost
Environmentally safe	Labor intensive
Low water absorption after installation	Difficult installation

to analyze application effectiveness and acceptability of the finished appearance.

Grouting is a highly labor intensive system, and the mechanics doing it should be trained and experienced in system application. Should grout be brushed off too quickly, material will be removed from masonry pores and will not sufficiently waterproof. If grout is allowed to cure completely, it will be virtually impossible to remove, and entire substrate aesthetics will be changed.

Application timing and removal varies greatly and is affected by weather (dry, humid, sunny, or overcast), substrate conditions (smooth, glazed, or porous), and material composition. Mechanics must be experienced to know when the removal process should begin, as this may change daily depending on specific project conditions, including weather (Table 7.6).

Face Grouting Application

Masonry walls should be cleaned completely to ensure that grout will bond to both existing masonry and mortar. All contaminants, including previously applied sealants, must be removed. Walls should be checked for residue of previous waterproof coatings or sealer applications that hinder grout bonding. All seriously deteriorated mortar joints should be tuck-pointed before grout application.

Grout materials are supplied in dry mix form with acrylic or integral bonding agents. Dry bag mix products are mixed with clean water in specified portions for existing conditions.

Grout should be brushed and scrubbed in circular motions to an entire wall area. The wall surface must be kept continually and uniformly damp to prevent grout drying before removal. Grout should be applied uniformly and completely to fill all voids, pores, and cracks (Fig. 7.9).

At the proper time, determined by job conditions, removal should begin. Grout is removed using stiff bristle brushes, burlap bags, or other effective methods. Proper removal will leave masonry free of grout deposits with no change in color or streaking.

Figure 7.9 Face or bag grouting application. (*Courtesy of Western Group*)

No priming is required, although surfaces should be kept properly damp. Materials should not be applied to unsound or defective substrates or joints. Temperature must be above freezing during application.

Joint Grouting

Joint grouting is an application of cementitious grout to all surfaces of existing mortar joints. This application is sometimes referred to as mask grouting, which is grouting walls that have all masonry units masked (taped or otherwise covered). This protects them from grout application on masonry unit faces. Materials used and surface preparation are the same as that for face grouting; only applications are different.

Cementitious grout material is brushed onto joint surfaces to fill voids and cracks, while keeping material off masonry facing. In restoration projects where joints have been tool recessed, grout application should fill joint recesses completely.

This application effectively points joints without requiring joint cutout. However, as with all joint grouting systems, severely deteriorated joints should be removed and properly tuck-pointed before grout application. (See Table 7.7.)

TABLE 7.7 Joint Grouting Properties

Advantages	Disadvantages
No aesthetic changes to substrate	Repairs only masonry joints
Less labor intensive than other methods	Adjacent surfaces should be masked
No damage to surrounding substrates	Joint removal required may be overlooked

Joint Grouting Application

If joints exist with a minor recess, ⅛ in or less, masonry units are masked and grout is applied to fill joints flush with the masonry facade. Masking is removed before complete curing of grout so that any fins formed may be removed before final grout is set without affecting waterproofing integrity (Fig. 7.10).

Figure 7.10 Mass grouting application. (*Courtesy of Western Group*)

This system is not designed to replace tuck-pointing seriously deteriorated joints. As with other systems, sample test areas should be completed to analyze effectiveness under specific job conditions.

Materials should be mixed according to the manufacturer's recommendations. Materials are brushed on existing mortar cracks and voids or applied by jointers to fill joint recesses completely. Grout materials are available in standard colors or are manufactured to match existing colors.

No priming is required, but joints should be kept damp during application. Materials should not be applied to frozen substrates or in freezing temperatures.

Epoxy Injection

During original construction or structure life cycling, cracks often develop that allow water and pollutants to enter into a substrate. If this cracking is nonmoving but structural, it is repaired through injection of a low-viscosity epoxy. The epoxy seals the cracks and restores the monolithic structural nature of a substrate.

High-strength epoxies can return a substrate to its original design strength but do not increase load-bearing capability. Epoxy used for injection has compressive strengths more than 5000 lb/in^2 when tested according to ASTM D-695.

Injection epoxies are two-component low-viscosity materials requiring mixing before application. Low viscosity allows materials to flow freely and penetrate completely into a crack area. Epoxy used for injection applications has no movement capabilities and will crack again if original cracking or movement causes are not alleviated. Expansion and control joints must be installed if it is determined that cracks may continue to move. Otherwise, cracks should be treated with a material that allows for movement.

Epoxy injection is a restoration system as well as a waterproofing system. Injection can restore substrates to a sound condition before waterproofing application or be used as waterproofing itself by stopping leakage through a crack.

Epoxy injection has been used on concrete, masonry, wood, metal, and natural stone substrates. Large wood timber trusses in historic structures have been restored structurally with epoxy injection systems. Typically, epoxy injection is used to restore concrete and masonry substrates to sound condition.

Cracks to be injected must be large enough to allow entrance of epoxy, approximately 5 mil thick, and not so large that material flows out, 35–40 mil. Cracks that meet these size limitations can be injected through any of the above listed substrate materials.

TABLE 7.8 Epoxy Injection Properties

Advantages	Disadvantages
Restores structural integrity	Extensive installation requirements
Wood, metal, and concrete substrate	No movement capability
High strength	May stain surrounding substrate

Application is completed by the pressure injection method using surface mounted or drilled ports through which to apply epoxy. In some cases, e.g., horizontal surfaces such as parking decks, epoxy is installed by the gravity method, in which epoxy simply penetrates by gravity. In all cases, a low-viscosity material is used to allow for better epoxy penetration into a substrate.

Surface mounted ports are applied directly over a crack surface. Drilled ports require a hole to be drilled at the crack location and a port placed into these holes for injection. Drilled ports are required for large deep cracks to allow complete saturation of cracks with epoxy.

In both port applications, cracks are sealed with a brushed-on epoxy to prevent epoxy from coming out of the crack face during injection. The port surround is also sealed and adhered completely to substrates, preventing them from blowing off during injection. If cracks penetrate completely through a substrate, the backside must also be sealed before injection.

The premise of injection work is to allow for maximum epoxy penetration to ensure complete joint sealing. Ports are placed approximately the same distance apart as crack depth, but not exceeding 6 in. Epoxy is then injected into the lowest port, and injection is continued until epoxy flows out the next highest port. The lower port is then sealed off, and injection is continued on next highest port. After epoxy curing, ports and surface applied sealers are removed.

Epoxy crack sealing will cause staining of a substrate or possible damage to substrates during its removal. In restoration procedures where such damage is not acceptable such as glazed terra cotta, hot-applied beeswax may be applied as a sealer in place of epoxy. This wax is then removed after injection without damage or staining to substrates.

Epoxy injection requires technical knowledge and experience of an installer. Proper mixing of the two-component materials, proper injection pressures, and knowledge of the injection process are mandatory for successful installations. (See Table 7.8.)

Epoxy Injection Application

Substrates must be cleaned and completely dry. If both sides of a substrate are accessible, they should both be sealed and injected to ensure complete crack filling. Cracks accessible from only one side of a substrate lose a quantity of material out to the unsealed side.

Cracks that are contaminated with dust or dirt cannot be properly injected. All cracks should be blown with compressed air to remove dirt accumulation. Steel substrates should be free of oxidation.

If epoxy is to be installed by the gravity method, horizontal cracks should be grooved to form a V-shape. The groove should be blown out to remove all concrete dust and other contaminants before epoxy is placed in the groove.

Cracks must be sealed with a brushable epoxy gel or wax to prevent epoxy run out during injection. Ports should be installed using either surface mounted or drilled ports as recommended by the manufacturer.

Injection should begin at the lowest port and be injected until epoxy is visible at the next higher port. After sealing and capping of the injected port, the injection is then moved to the next port. Upon completion of port injection, substrate sealers and ports are removed.

Most epoxy used for injection is two component and must be properly mixed before application. Working life or pot life is extremely limited, and epoxy must be installed before it begins to cure in the applicator equipment. Epoxy injection equipment that mixes and injects epoxy under constant uniform pressure, approximately 100–300 lb/in^2, is available. Low-viscosity epoxy thickens in cool weather and may not flow sufficiently to fill the crack. Hand pump injectors are adequate as long as enough ports are used.

Upon injection completion, a core sample of substrate and installed epoxy should be taken. This allows for inspecting penetration depth and testing strength of cured epoxy and repaired substrate. Epoxy materials are extremely hazardous and flammable. Care should be taken during their use as well as their disposal. Equipment must be checked frequently to ensure that proper mixing ratios are being maintained.

Chemical Grout Injection

Epoxy materials are used for restoring substrates to sound structural strength, with waterproofing of cracks a secondary benefit. Epoxy joints do not allow for movement. If movement should occur again, leakage can resume. Chemical injection grouts, on the other hand, are used primarily for waterproofing a substrate and are not intended for structural repair. Chemical grouts also allow for future movement at joint locations.

Injection grouts are hydrophobic liquid polymer resins such as polyurethane formulations. They react with water present in a crack and substrate, creating a chemical reaction. This reaction causes a liquid grout to expand and form a gel or foam material that fills voids and

TABLE 7.9 Chemical Grout Properties

Advantages	Disadvantages
Excellent movement capability	Cost
Several formulations available	Requires specialized installation methods
Concrete, masonry, wood, and soil substrates	Toxic chemical used

cracks. Expansion of materials form a tight impervious seal against substrate sides, stopping water access through a joint.

Grout material is supplied in low-viscosity formulations to enhance its penetrating capabilities. However, unlike epoxy, substrates do not need to be dry for grout application. In fact, substrates should be wetted before application and, if necessary, grout can be applied directly into actively leaking joints.

Grouts are typically used for concrete or masonry substrates, although they will bond to metals, wood, and polyvinyl chlorides; such as PVC piping. Some grouts are also available in gel form, which is used to stabilize soils in areas of bulkheads or soil banks of retention ponds. In these applications, materials react with groundwater present, binding together soil particles.

Grout injected into wetted substrates fills fissures and pores along a crack surface. Once-cured grouts, similar to sealants, have excellent movement capability with elongation as much as 750 percent. This flexibility allows material to withstand thermal movement or structural movement at a joint without deterring its waterproofing capabilities. These grouts have been used successfully in remedial below-grade applications where leakage is occurring directly through a crack itself rather than through entire substrates.

Since water moves through a path of least resistance, during remedial repairs injecting cracks can redirect water and start leakage in other areas of least resistance. Therefore, complete remedial waterproofing treatments can require grout injection of cracks and application of a waterproofing system. (See Table 7.9.)

Chemical Grout Application

Chemical grout applications are very similar to epoxy injection, with similar equipment and injection tubes necessary. The major difference is that grouts require water and epoxies do not. Additionally, chemical grouts are supplied in one component rather than two-component epoxy formulations.

Substrate preparation is almost unnecessary with chemical grouts. Surfaces do not need to be dry but should be cleaned of mineral deposits or other contaminants along a crack area. If a waterproofing system is to be applied after grout injection, the entire substrate should be cleaned and prepared as necessary.

On minor cracks, or substrate of 6-in thick or less, holes for ports are drilled directly over cracks. In thicker substrates and large cracks, ports are drilled approximately 4–6 in away from cracks at an angle that intersects the crack itself. Test holes should be completed with water injected for testing to ensure that grout will penetrate properly. Port spacing along cracks varies, depending on crack size and manufacturer recommendations. Spacing varies from 6 to more than 24 in.

In smaller cracks it is not necessary to surface seal crack faces. However, on large cracks, temporary surface sealing with a hydraulic cement patching compound is necessary to prevent unnecessary grout material waste. As with epoxy, applications begin at lowest port, and grout is injected until it becomes present at the next higher port. This process is then moved to the next higher port.

Pressures required to inject materials are generally 300–500 lb/in^2. After injection, material should completely cure before injection ports are removed, approximately 24 hours. Port holes should then be patched with quick set hydraulic patching material.

Chemical grouts should not be used in temperatures below 40°F nor on frozen substrates. Chemical grout materials are flammable and hazardous. Extreme care should be taken during its use and storage as well as the disposal of the chemical waste. In confined spaces, ventilation and respirators are required for safe working conditions.

Cementitious Patching Compounds

For restoration of concrete and masonry substrates, a host of cementitious-based products is available to restore substrates to a sound condition before remedial waterproofing applications. These cement-based products are high strength, dry mixes, with integrally mixed bonding agents or a bonding agent to be added during mixing. The premixed products have similar properties: they are high strength (compressive strength usually exceeds 5000 lb/in^2), fast setting (initial set in less than 30 minutes), applicable to damp substrates, and nonshrinkable.

Some products are used as negative side waterproofing systems, whereas others are used to patch a specific area of leakage. Cementitious materials are used in a variety of restoration procedures. A typical use is repairing spalled concrete surfaces after repair and preparation of exposed reinforcing steel.

Cementitious patching systems are also used for concrete overlays to add substrate strength, patching of honeycomb and other voids, and patching to stop direct water infiltration. Cementitious patching systems include

- High-strength patching compounds
- Hydraulic or hot-patch systems
- Shotcrete or gunite systems
- Overlays

High-strength patching

High-strength patching products are supplied in premixed formulations with integral bonding agents. They are used to patch spalled areas or voids in concrete or masonry to prepare for waterproofing application. These products contain a variety of proprietary chemicals and additives to enhance cure time, strength, and shrinkage. High-strength patching systems require only water for mixing and can be applied in a dry or stiff mix that allows for vertical patching applications.

Properties of cementitious high-strength mixes vary, and product literature should be reviewed to make selections to meet specific repair conditions. Most important, installation procedure with all products is maximum single-application thickness.

Most products require maximum 1-in layers due to the chemical process that creates extreme temperatures during curing. If an application is too thick, patches will disbond or blow off during curing.

Hydraulic cement products

Hydraulic products are frequently referred to as hot patches because of heat generated during the curing process. Hydraulic refers to running water, such as water leakage through a crack. These materials set in an extremely short time due to this internal chemical curing, which dries the material rapidly. This property enables these materials to patch cracks that exhibit running water in concrete or masonry.

To complete repairs to substrates, leaking cracks are sawn out, approximately 1 × 1 in. This groove is then packed with hydraulic cement, regardless of any running water present. The material sets in approximately 5 minutes, in which time leakage is effectively sealed. Should water pressure be great enough to force patch material out before initial set, relief holes must be drilled along the crack to redirect water. These holes will continue to relieve water pressure until the crack patching has cured, after which relief holes themselves are patched to stop water infiltration completely.

These products are also used to seal port holes in epoxy and chemical grout injection. Hydraulic materials are often used in conjunction with negative cementitious waterproofing applications to complete substrate patching before waterproofing material application.

The extremely fast initial set prohibits their use over large wall or floor areas. They are limited to patching cracks or spalled areas that can be completed within their short pot life.

Shotcrete or gunite

Shotcrete and gunite refer to pneumatically applied small aggregate concrete or sand–cement mixtures. These are used to restore existing masonry or concrete substrates to a sound structural condition, waterproofing preparation, or both. These methods are used when areas requiring restoration are sufficiently large, making hand application inefficient. Pumping and spray equipment used in shotcrete can automatically mix materials, then pneumatically apply them to substrates.

Gunite or shotcrete mixtures used with this equipment vary from field ratio mixes to premixed manufactured dry materials requiring only the addition of water. Materials are applied as a dry mix for vertical applications. After initial application, materials are troweled or finished in place as necessary.

Surface preparation requires chipping and removal of all unsound substrates areas and repairing of existing reinforcing steel as necessary. Additional reinforcing steel may also be installed if necessary before gunite operations.

Overlays

Cementitious overlays are used for restoring deteriorated horizontal concrete substrates. Overlays are available in a wide range of mixtures containing various admixtures that add strength and shorten curing time. They are used in a variety of applications including bridge repairs and parking deck restoration.

Often they are sufficiently watertight to eliminate a need for waterproofing coatings. Others are designed specifically for use as an underlay for deck coating applications. If additional structural strength is necessary, qualified engineers should be consulted for selection and use of such products.

These materials are usually self-leveling, conforming to existing deck contours to which they are applied. They are used also to fill ponding or low areas of existing decks before deck coating application. Stiffer mixes are available for ramp areas and inclined areas. (See Table 7.10.)

TABLE 7.10 Cementitious Systems Properties

Advantages	Disadvantages
Variety of systems available	No movement capability
Negative or positive systems	Only masonry and concrete substrates
Large or small repair areas	Mixing controlled at site

Chapter

8

The Building Envelope: Putting It All Together

The entire exterior facade or building skin must be completely watertight to protect interior areas, maintain environmental conditioning, prevent structural damage, and provide economical life cycling. It is only with exacting attention to details by designers, installers, and maintenance personnel alike that building envelopes maintain their effectiveness.

Too often, items are specified or installed without adequate thought as to how they will affect envelope performance and whether they will act cohesively with other envelope components. For example, rooftop mechanical equipment must by itself be waterproof, but connections attaching the equipment to the building envelope must also be waterproof.

Rooftop mechanical equipment relies on sheet metal flashing, gasketed closure pieces, and drip pans for transitions and terminations into other envelope components. Once installed, these details must act cohesively to remain watertight.

Little thought is given to the performance of these transitions during weathering, movement, and life cycling. This results in leakage and damage, which could be prevented by proper design, installation, and maintenance. Envelope portions that are often neglected and not made watertight include lightning equipment, building signs, vents, louvers, screens, heating/ventilating/air conditioning (HVAC) and electrical equipment, lighting fixtures, doorways, and thresholds.

Envelope Waterproofing

The first step in designing new envelopes or reviewing existing envelopes is to ensure that all major facade components are waterproof, acting as first-line barriers against water infiltration. Brick, glass, metal, roofing, and concrete must be waterproof; otherwise, allowances must be incorporated to redirect water that bypasses these components back out to the exterior.

Considering masonry wall construction, it is typically not the brick that allows water to enter but the mortar joints between the brick. One square foot of common brick wall area contains more than 7 linear feet of mortar joints. These field-constructed joints are subject to installation inconsistencies that occur with site construction.

It takes but a 1/100-in crack or mortar disbonding for water infiltration to occur. This cracking occurs as a result of shrinkage of the mortar, settlement, differential movement, wind loading, and freeze–thaw cycles. Multiply these 7 lineal feet by the total brick area, and the magnitude of problems that might occur becomes evident.

To offset this situation, brick joints should be expertly crafted, including properly mixing mortar, using full bed joints and proper tooling of joints. Joint toolers compress mortar against both sides of attached brick, compacting the material, which assists in preventing water from passing directly through joints.

Water that passes through joints carries salts extracted from cement content in mortar. The whitish film often occurring on exposed masonry walls is referred to as efflorescence. It is formed by salt crystallization after water carrying the salts is drawn by the sun to the surface. This water then evaporates, leaving behind the salt film.

When salts crystallize within masonry pores, the process is called cryptoflorescence. Formation of these crystals can entrap moisture into masonry pores that cause spalling during freeze–thaw cycles. Additionally, if cryptoflorescence is severe enough, it will prevent the natural breathing properties of a masonry wall. Both forms of salt can attack and corrode reinforcing and supporting steel, including shelf angles. This corrosion often leads to structural damage.

An effective joint tooling method is a weathered joint finishing. In this tooling installation, a diagonal is formed with mortar, with the recess at top, allowing it to shed water. Recessed joints, including struck and raked joints, can accumulate water on horizontal portions of the recess and exposed brick lip. This water may find its way into a structure through mortar cracks and voids. Commonly used joint types appear in Fig. 8.1. All masonry mortar joints as well as all building envelope components should be designed to shed water as quickly as possible.

Once major envelope components are selected and designed, transition systems are chosen to detail junctures and terminations of the major components. Transition materials and systems ensure the watertight integrity of an envelope where changes in facade components occur or at terminations of these components.

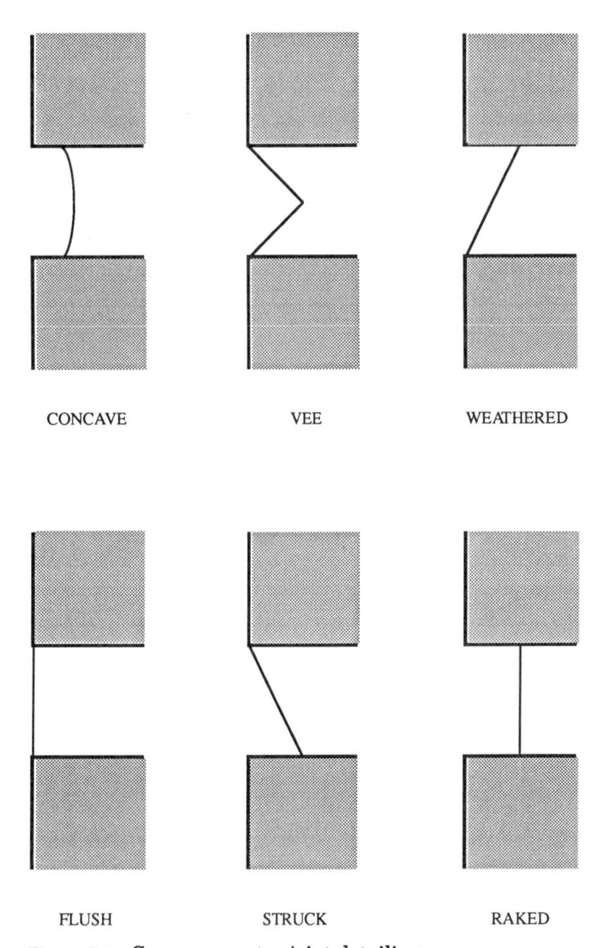

Figure 8.1 Common mortar joint detailing.

Transition Materials

There are two basic cladding details for above-grade envelope facades used to prevent water infiltration. The first type is a solid, single-barrier wall system with no backups or secondary means of protection (e.g., single wythe block walls, stucco over metal lath, and exterior insulated finish systems). The second type is a multicomponent system providing at least two waterproofing methods. These include the cladding itself and backup systems consisting of flashing and weeps that redirect water passing through first-line barriers (e.g., brick cavity wall, metal and glass curtain walls with integral gutters, flashing, and weeps).

Multicomponent systems provide better resistance to water infiltration by providing systems that redirect water bypassing the initial barrier back out to the exterior. These systems are most effective when this redirecting is channeled immediately out to the exterior and not allowed to drain into other interior systems. The latter allows an envelope to become susceptible to leakage into interior areas.

As previously emphasized, however, it is typically not the primary waterproofing barriers themselves that directly cause water leakage, rather the transitions between these envelope components. Transitions and detailing between major components create 90 percent of leakage problems, although they represent only 1 percent of an envelope area. Besides permitting leaking water into interior areas, this leakage creates damage to structural components.

Manufacturers provide recommended installation details for their products and systems, and these should be followed without exception. If a particular installation presents specific detailing problems, a manufacturer's representative will review the proposed transitions and actual installations as necessary. This inspection requirement is one of the major advantages of joint manufacturer and contractor warranties discussed in Chap. 9.

In addition to manufactured system components for transitions, several frequently used transition systems and materials are used in building construction. These systems provide watertight transitions between various primary envelope components when installed properly. Standard available systems include

- Flashings
- Dampproofing
- Sealant joints
- Reglets
- Waterstops
- Pitchpans
- Thresholds
- Expansion joints
- Cants

Any of these standard systems proposed to be used as part of a waterproofing or envelope system should first be reviewed and approved by the waterproofing system manufacturer. This eliminates unnecessary problems that prohibit envelope components acting cohesively together and preventing water infiltration.

Flashings

Because water is likely to pass through masonry facades, cavity wall construction with dampproofing and flashing systems are necessary to redirect entering water. Dampproofing materials, usually asphaltic or cementitious compounds, are applied to the outer faces of interior wythes. This prevents minor amounts of water or moisture vapor from entering interior spaces. Dampproofing requires flashing to divert accumulated water and vapor back to the exterior through weep holes.

Envelopes often depend on flashing for maintaining watertight integrity. Flashings are not only used in brick masonry veneer structures but also in the following:

- Poured-in-place concrete
- Precast concrete panel construction
- Stucco or plaster veneer walls
- Insulated wall systems
- Stone veneers
- Curtain and window wall systems

Flashings are manufactured from a variety of materials, including noncorrosive metals and synthetic rubber sheet goods. Metal flashings include copper, aluminum, stainless steel, galvanized steel, zinc, and lead. Sheet-good flashings are usually a neoprene rubber or a rubber derivative.

Thermal expansion and contraction that occurs in a facade also introduce movement and stress into flashing systems. If installed flashing has no movement capability, it will rupture or split, allowing water infiltration. Adequate provisions for thermal movement, structural settlement, shear movement, and differential movement are provided with all flashing systems.

Flashing installed in cavity walls typically is the responsibility of masonry contractors, who often do not realize the importance of properly installing flashings to ensure envelope effectiveness. Common flashing installation problems include

- Seams not properly spliced or sealed
- Inside or outside corners not properly molded
- Flashing not meeting at building corners
- Flashing improperly adhered to substrates
- Flashing not properly shedding water

Besides these problems, masons often fill cavities with mortar droppings or allow mortar on flashing surfaces. Additionally, weep holes are filled with mortar, damming water from exiting.

These examples, and all problems associated with site construction,

TABLE 8.1 Commonly Used Flashing Systems and Their Functions

Location	Function
Base flashing	Prevents capillary action of water from wicking upward in a masonry wall
Sill flashing	Installed beneath window or curtain wall sills
Head flashing	Installed above window head detail, just below adjacent facing material that the window abuts
Floor flashing	Used in conjunction with shelf angles supporting brick or other facade materials
Parapet flashing	Installed at the parapet base, usually at ceiling level; may be used on roof side of parapets as part of roof or counterflashing
Counterflashings	Surface mounted or placed directly into walls with a portion exposed to flash various building elements into the envelope, including roof flashings, waterproofing materials, building protrusions, and mechanical equipment
Exposed flashings	Used in a variety of methods and locations; can be an integral part of an envelope system, such as skylight construction, or applied to provide protection between two dissimilar materials, including cap flashings, coping flashings, gravel stops, and edge flashings
Remedial flashings	Typically surface mounted and applied directly to exposed substrate faces; can include a surface-mounted reglet for attachment; do not provide for redirecting entering water; only by dismantling a wall or portion thereof can remedial through-wall flashings be installed

make it necessary for all subcontractors to be made aware of their responsibility and of the interaction of all building envelope components. Making frequent inspections during all envelope work ensures quality, watertightness, and cohesiveness. As with all envelope components, attention to flashing details, only 1 precent of building area, ensures the success of waterproofing envelopes.

Flashings should be installed at vulnerable areas in envelopes and where necessary to redirect entering water. Typical flashing locations and their basic functions are summarized in Table 8.1.

Flashing Installation

Flashing applied to concrete substrates requires that the concrete be clean, cured, and free of all honeycomb, fins, and protrusions that can puncture flashing materials. If applied by mastic, substrates should be clean and dried. Mechanically attached systems require that substrates be sound, so as to allow anchoring of attachments. Flashings should extend up vertically at least 12 in from the horizontal installation point.

Flashings set on the top of shelf angles should only be manufactured of materials that do not cause galvanic reaction with the steel angles. Often the horizontal joints along shelf angles function as control joints. If sealant is installed in this joint, it is applied below the shelf angle and flashing to prevent interference with water exiting the envelope.

Exposed flashing such as roof, cap, or coping is installed to provide a transition between dissimilar materials or protection of termination details. These flashings should be securely attached to structural elements to provide resistance against wind loading. Only seams that allow for structural movement, usually as a slip joint design, should be used.

All seams in both metal and sheet materials must be properly lapped and sealed. Flashing systems allow water infiltration because of improper attention to seams, bends, and turn details. Often flashings and shelf angles are inadvertently eliminated at building corners, allowing a continuity break and a path for water to enter.

Flashing must have adequate provisions to allow for thermal and differential movement as well as for shear or deflection of wall areas, not only for longitudinal movement but for vertical movement and shear action occurring when inner walls remain stationary while outer walls experience movement.

Detailings of flashings intersecting expansion joints in exterior wall systems are also likely failure areas. There is usually a break in structural framework at these locations, allowing for structural movement. At these locations, flashings are terminated, with their ends dammed and detailed to allow for this movement. Typical flashing installations in common building materials are shown in Fig. 8.2.

Dampproofing

Dampproofing materials are typically used in conjunction with flashing and weep systems as part of secondary or backup systems for primary envelope waterproofing materials such as masonry walls. In addition, dampproofing systems are used at below-grade applications to prevent moisture vapor transmission or capillary action through concrete or masonry walls. Dampproofing can be applied in either negative or positive installations.

Dampproofing prevents damage to envelope components where surface water can collect and drain to below-grade areas. It also protects when improper surface water collection systems are not used in conjunction with below-grade drainage mats.

Dampproofing materials, as defined, are not intended to, nor do they,

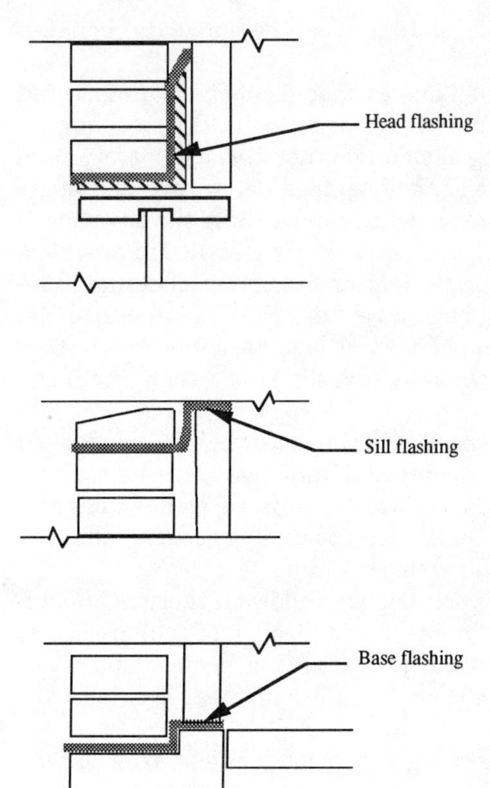

Figure 8.2 Envelope flashing detailing.

function as primary envelope waterproofing systems. They function only as additional protection for the primary envelope systems or are used where water vapor or minor amounts of water are expected to be encountered on an envelope. They are used in below-grade applications when no hydrostatic pressure is expected or when it will occur in only the severest of expected weathering.

Dampproofing systems are available in a variety of materials and systems, including

- Cementitious systems
- Sheet vapor barriers
- Bituminous dampproofing

Clear water repellents are also sometimes referred to as dampproofing materials because they are not effective against hydrostatic water pressure. However, clear sealers are typically applied directly to the face of primary envelope waterproofing materials or facades. As such they do not function as dampproofing materials or backup and secondary systems. Clear water repellents are discussed in detail in Chap. 3.

Cementitious systems

Cementitious dampproofing systems, which are available in a wide range of compositions, are sometimes referred to as *parge coats*. Parging is an application of a cementitious material applied by trowel to a masonry or concrete surface for dampproofing purposes. Parging is also used to provide a smooth surface to substrates before waterproofing material application.

Cementitious dampproofing materials are usually supplied in a dry premixed form. They are cementitious in a base containing the manufacturer's proprietary water-repellent admixture. Mixes also include bonding agents. Their bonding capability to masonry or concrete substrates provides an advantage over sheet materials or bituminous materials, since cementitious materials become an integral part of a masonry substrate after curing.

In addition, cementitious systems can be applied to damp substrates in both above- and below-grade positive or negative locations. An example is application to the interior of elevator pits not subject to hydrostatic pressure.

Cementitious applications also prevent capillary action at masonry walls and foundations or floor slabs placed on soils that are subject to capillary action. Dampproofing in these areas prevents water vapor transmission to upper and interior envelope areas that can cause damage, including deterioration of flooring and wall finishes.

Sheet or roll goods

Sheet dampproofing materials are manufactured from polyvinyl chloride, polyethylene, and combinations of reinforced waterproof paper and polyvinyl chloride. They are available in thicknesses ranging from 5 to 60 mil. Sheet materials are typically used for dampproofing horizontal slab-on-grade applications to prevent capillary action through floor slabs.

Sheet systems have limited vertical application uses. They are used to wrap shallow below-grade pits when waterproofing materials are not necessary. Sheet materials are typically not used in above-grade vertical areas because of the difficult application methods at these areas.

Sheet systems are more difficult to transition into other envelope waterproofing systems, particularly at below-grade to above-grade

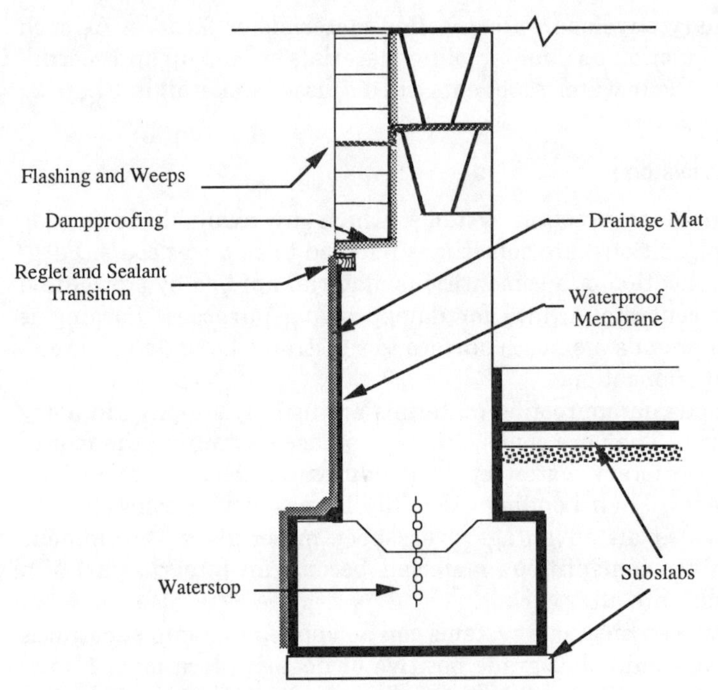

Figure 8.3 Dampproofing to waterproofing transition.

transitions. Typically, at these areas a mastic material is used to adhere the sheet material and provide a transition to the above-grade materials. Refer to Fig. 8.3 for a typical detail for above-grade mastic dampproofing material transitioning into slab-on-grade sheet materials.

Bituminous dampproofing

Bituminous dampproofing materials are either asphaltic or coal-tar pitch derivatives. They are available in both hot-applied and cold-applied systems, with or without fiber reinforcing. Coal-tar derivatives are seldom used today due to health risks and safety concerns during installation.

Glass or fabric fibers are added to dampproofing materials that allow trowel or brushable applications by binding the material together in a thicker consistency. Reinforcement also adds minor durability characteristics to the material but not to water-repellency capabilities.

Asphaltic products are available in an emulsion formulation (water based). Besides allowing easier applications and cleanup, water-based dampproofing materials are breathable, allowing vapor transmission in envelope areas, such as parapet wall applications, where this is necessary.

Hot-applied systems

Both hot asphaltic and coal-tar pitch systems are typically used for below-grade positive applications. Difficulties involved in installation prohibit most interior (negative) applications.

Materials used in hot-applied systems are typically those used in built-up roofing applications with the addition of roofing felts. These are usually applied in a one-coat application.

Difficulties in installations, equipment required, and field quality control has greatly limited hot-applied system usage. Cold-applied dampproofing systems that meet or exceed the performance of hot-applied systems are available, including those incorporating fibrous reinforcement.

Cold-applied systems

Cold-applied dampproofing systems are available in both coal-tar and asphaltic-based compositions. These systems are solvent-based derivatives, with or without fibrous reinforcement, that cure to form seamless applications after installation. Unfibered or minimal fiber systems are applicable by spraying. Heavily reinforced systems materials are applicable by trowel or brush.

Typically used on concrete or masonry substrates, cold-applied dampproofing materials can also be used on metal, wood, and natural stone substrates. Cold systems are used in both positive and negative applications, both above and below grade. Negative systems are applied to walls that are furred and covered with drywall or lath and plaster.

Negative systems do not allow for the collecting and redirecting of water entering an envelope. Therefore, negative applications are used only when vapor transmission through the primary waterproofing barrier is expected.

Cold-applied emulsion-based asphalt systems are also available. These water-based systems offer easy cleanup and are used where solvent systems can damage adjacent flashings, waterproofing materials, or substrates themselves. Some cold-applied emulsion-based systems are applicable over slightly damp or uncured concrete, allowing for immediate dampproofing after concrete placement.

Emulsion systems should not be used in any below-grade applications or above-grade locations where sufficient amounts of water are present that can actually wash away the dampproofing material from a substrate. In addition, emulsion-based systems must be protected from rain immediately after installation. This protection must be adequate to keep installations protected until primary envelope materials are in place and backfill operations are complete.

Dampproofing Installation

Dampproofing applied to concrete or masonry substrates requires surfaces to be clean, cured, and free of all honeycomb, fins, and protrusions. Some emulsion systems allow application over uncured or slightly damp concrete. Sheet materials applied directly over soil should be placed on compacted and level granular soils that do not promote capillary action.

Negative applications are used when only vapor transmission is expected through primary envelope waterproofing systems. If water is expected to enter through the primary envelope components, negative systems should not be used. Negative systems provide no means to collect and redirect this water to the exterior.

Positive systems are used in conjunction with flashing and weeps to redirect entering water to the exterior. Water-based systems should not be used where substantial amounts of water are expected to enter and collect on the dampproofing, as this can wash the water-based material off the wall, particularly in below-grade construction.

Mastic applications are applied in thicknesses ranging from 30–35 mil. Sheet systems are generally 10–20 mil thick. Cementitious systems are troweled applied to thicknesses of approximately ⅛ in. Millage applications should be checked regularly to ensure that proper thicknesses are being applied.

When applying dampproofing to inner whytes of masonry veneer walls incorporating brick ties, a spray application of mastic is most suitable. Spraying allows for a uniform coverage around the ties, which is difficult using a trowel.

Dampproofing used in conjunction with flashing systems should be installed after flashings are adhered to the substrate. Applying dampproofing after the flashing fasteners are in place is preferable to having the dampproofing punctured during flashing application.

The dampproofing used should be compatible with flashing materials. Some solvent materials can damage sheet flashing systems. Dampproofing should extend over the flashing and attachments to allow adequate transition detailing and ensure proper drainage of water onto the flashing where it can be redirected to the exterior. Refer to Fig. 8.4 for a typical flashing and dampproofing transition.

For negative installation, first, furring strips should be installed, then, dampproofing materials should be installed. This prevents dam-

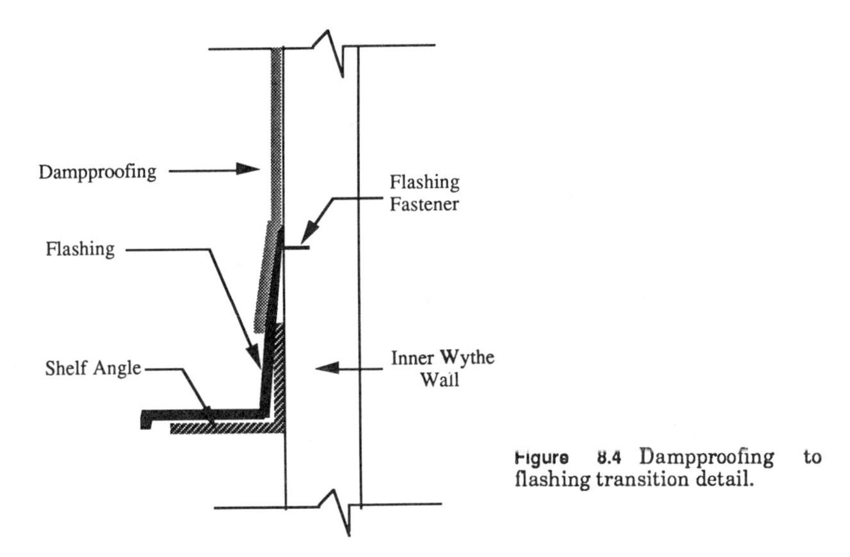

Figure 8.4 Dampproofing to flashing transition detail.

age of the dampproofing continuity by fasteners used for attaching the furring strips. With cementitious negative systems, furring strips can be directly applied with adhesives to the dampproofing to prevent damage.

Sealant Joints

Sealant materials are frequently useful in providing transitions between dissimilar materials or systems in a building envelope. They also provide watertight allowances for thermal or dissimilar movement between these components. For instance, joints between metal window frames and wall facades provide a watertight transition between window and wall components and allow thermal and differential movement.

Sealants are often overused, especially in remedial waterproofing repairs. Simply applying sealants over failed areas will not adequately address failure problems. A thorough investigation or study should first be completed to determine why materials or systems originally failed.

For example, sealants are often used in place of tuck-pointing to correct mortar joints. This is a poor approach if existing mortar is not of sufficient strength to maintain envelope integrity. Such applications also allow three-point adhesion and improper sealant depth, and under structural movement sealant installation repairs will fail.

Control joints provide terminations or transitions that use sealants to seal and waterproof a joint. Common control joints include joints between dissimilar products, transitions between vertical and horizontal junctures, and equipment protrusions such as plumbing, electrical piping, lighting equipment, and sign supports.

Precast panel construction with porous finishes allows water transmission directly through the panels, bypassing joint sealants. Absorption of water through panels is further enhanced by negative air pressure between the exterior and interior areas. This uneven air pressure causes water to be drawn into a structure by a suction process. Likewise, wind-driven rain forces water through minor cracks and fissures in a masonry or concrete structure.

These natural phenomena are addressed by double sealing joints with sealant being applied to both exterior and interior sides of panelized construction. This double sealing allows air pressure in wall cavities to remain relatively constant by pressure equalizing the sealed space between exterior and interior areas. Double sealing also provides additional air seal protection for buildings, reducing heating and cooling costs.

With cavity wall construction, water enters through initial weather barriers (e.g., brick facing) and is redirected to the exterior. Water entering in such conditions reacts with alkalines in masonry, causing a highly alkaline solution that deteriorates all types of sealant. This causes sealants to reemulsify and leads to adhesion failure.

Therefore, weeps in masonry walls and at sealant joints must be kept clear and working effectively to prevent damage to sealant materials. This is accomplished by installing a plastic weep tube at the bottom of each sealant joint.

Reglets

Reglets are also used to provide for transitions or terminations in materials or systems of building envelopes. Reglets are small grooves or blockouts in substrates. Materials are turned into these reglets to be terminated or allow transitions between two different materials. Reglet uses within building envelopes include

- Substrate termination
- Waterproofing material to substrate transition
- Waterproofing system to waterproofing system transition
- Waterproofing system termination

Waterproofing materials that run vertically up from a foundation wall often terminate in reglets above grade. Below-grade waterproofing materials, changing to dampproofing systems above grade, use reglets for transitions between these two systems.

Reglets are formed in substrates during the placement of concrete by using blockouts. They are also formed by sawing concrete, masonry, or wood substrates to form a reglet recess. After a reglet is in place, it is inspected for cracking, honeycomb, or other problems that can cause leakage.

With certain flashing systems, surface-mounted reglets, mechanically fastened to substrates are used. These are often used in remedial waterproofing repairs or roofing installations where existing reglets are not functioning or do not exist. Reglets are not recesses placed into a substrate for crack control, such as recessed control joints. These control joints are later sealed or provided for aesthetic purposes, as in precast panel construction.

Waterstops

Waterstops, although limited to concrete construction, are highly effective for transitions between separate concrete pours (referred to as cold joints) and terminations between vertical and horizontal concrete placements. Waterstops are now produced in a wide variety of designs and materials including extruded rubber and hydros clay materials.

Remedial waterstop applications are now available. Cold joints with or without failed waterstops are chipped or sawn out along cracked areas. Manufactured tubing is then placed into this chipped-out area, and the joint is packed with nonshrink grout and cured. Then a urethane grout material is pumped into the tubing and filled, forcing expansion of the tubing and effectively sealing the joint.

Waterstops are effective in preventing lateral movement of water at cold joint areas. These joints are subject to cracking due to concrete shrinkage and allow water penetration if a barrier such as a waterstop is not installed.

Waterstop materials are manufactured with flanges that allow each side of a joint to be securely anchored to the concrete. A common problem with waterstop usage is that waterstops are frequently installed by concrete finishers who do not understand their importance and effect on a building envelope.

Waterstops often end up bent over, cut, or not lapped properly at seams. They are even completely removed during concrete placement because they get in the way of the concrete finishing process. Waterstop installations must be carefully inspected by construction management personnel during concrete placement operations to ensure such activities do not occur.

Other Transition Systems

A variety of other transition materials and systems is frequently used in construction practices to provide complete waterproofing of the building envelope. Among them are

- Pitch pans
- Thresholds
- Integral flashings of curtain and window wall systems
- Expansion joint systems
- Cants

All of these systems ensure the envelope watertightness by providing a transition or termination between dissimilar materials. They also allow for differential movement between various waterproofing systems or allow entering water to be redirected to the exterior.

For instance, thresholds placed under doorways prevent rain and wind from entering into interior spaces. They also provide a watertight transition among the exterior surface, doorway, and interior areas. Pitch pans provide a watertight transition among mechanical equipment supports, roofing, and structural roofing components.

It is the lack of such systems or inattention to proper application of these transitory materials that leads to most leakage in an envelope. It is the 1 percent of detail problems that causes 90 percent of water infiltration problems.

After all major envelope components and their transitions and termination details are complete, a review of the total envelope is made to ensure it will act cohesively. This review should begin with ensuring that all types of water reaching an envelope drains away quickly. This prevents unnecessary infiltration and weathering of envelope components.

Drainage Review

An important point in reviewing any particular building envelope, existing or in design, is that water should be shed and removed away from a building as quickly as possible. From below grade to roof, surfaces should be sloped whenever possible to shed water quickly.

Drainage must be provided to remove this water from surrounding areas. This not only prevents water leakage but also prevents premature weathering or wear of building envelopes from such sources as acid rain, chloride contamination, algae attack, and standing or ponding water. Refer to Fig. 8.5 for recommended drainage requirements.

Walls and floor areas below grade are waterproofed to form a complete monolithic enclosure. Structures are subjected to water not only from groundwater conditions but also from run-off from surface collected water. Buildings built at or below-grade level are also subject to a head of water. They are also exposed to water from capillary rise from being in contact with the ground and water percolating downward from above-grade sources.

Below-grade surfaces typically have the most severe conditions with which to contend, with the exception of wind-driven rain that presents water pressures higher than a head of water. Particular attention to all of these conditions is necessary to ensure integrity of below-grade envelopes.

Existing soil conditions also require consideration. Below-grade horizontal surfaces should be placed on top of coarse granular soil (e.g.,

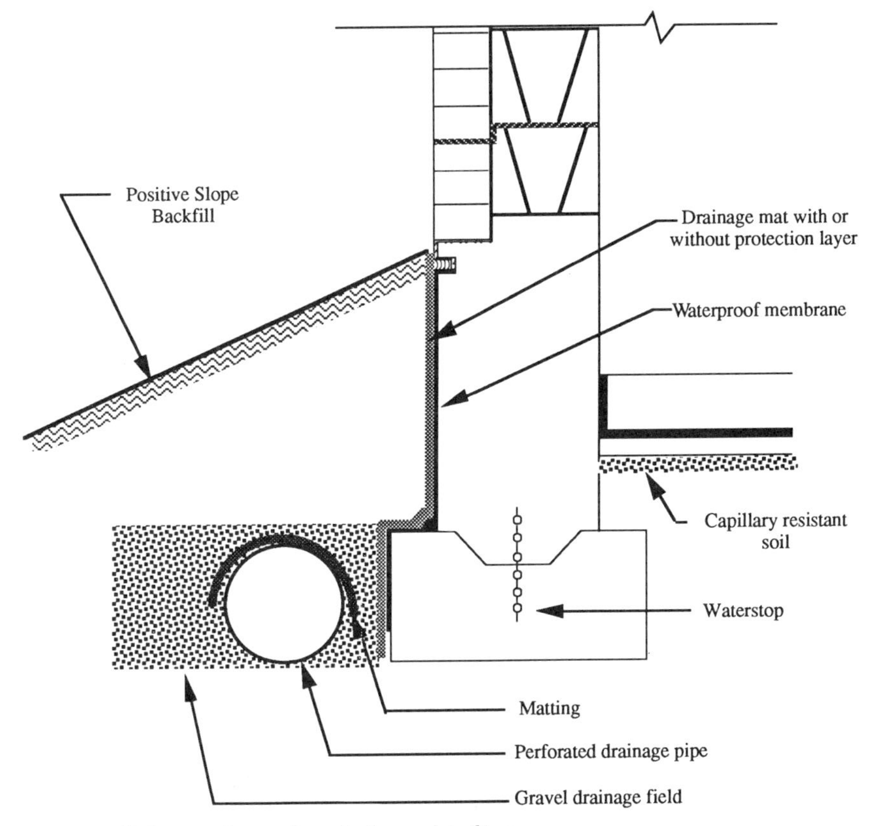

Figure 8.5 Below-grade envelope drainage detailing.

sand, gravel) to allow positive drainage away from foundations. Granular soils also resist capillary water rise that occurs in dense soil materials such as clay or silts. Areas adjacent to wall surfaces below grade should consist of coarse granular material to promote drainage.

Proper drainage of subsurface water is a necessity for adequate ensurance against water infiltration and longer life cycling of building components below grade. Foundation drains are installed as shown in Fig. 8.5. Drains are placed at foundation level or slightly above to prevent washout of soil beneath foundation structures. Drains are usually perforated piping with holes facing downward so as not to fill the pipe with soil and are set in a coarse granular bed. They should be sloped away from building structures, with water collected at a natural outface or sump area such as a retention pond.

It is also recommended that drainage mats be installed on vertical surfaces of waterproofing membranes below grade. Several synthetic compositions are available and compatible with waterproofing membrane systems. These mats promote quick drainage of water off below-grade walls into available drainage systems.

At grade level, grading should be sloped away from structures to provide positive drainage of surface waters away from buildings. Slope ratios differ depending on the type of soil and adjacent exposed finishes at surface level. For planted or grassed areas, slopes should be 5 percent minimum. For paved areas 1 percent minimum slope is acceptable. Draining water is collected and properly diverted to prevent excess water percolating into soil adjacent to structures.

Walls above grade are subjected to water from rain, snow, and capillary action of soil at grade level. Water conditions that are present can become especially severe when rain is present in high wind conditions, forcing water through minute cracks and openings in above-grade envelopes. Wall areas must also withstand weathering conditions to which a below-grade envelope is not subjected. These conditions include ultraviolet degradation, air pollutants, acid rain, chloride attack, freeze–thaw cycling, and thermal shock. Therefore, to be completely effective, exposed portions of envelopes must not only be watertight but also weather resistant. This ensures longevity of these systems and protection of interior areas.

Above-grade envelopes must also be provided with provisions to drain water away from a structure adequately, not allowing it to percolate down to below-grade areas. All building horizontal portions should be sloped to shed water. This includes not only roof areas but also coping caps, sills, overhangs, ledges, balconies, decks, and walkways.

To allow areas of standing or ponding water not only makes an envelope subject to water infiltration but also intensifies weathering from such sources as acid rain, chlorides, and algae. Where applicable, deck drains, roof drains, gutters, and downspouts should be installed to gather collected water and disperse it without subjecting above- and below-grade areas to this water.

Envelope Review

Successful building envelopes include most if not all of the following features:

- Few protrusions and penetrations on exposed envelope portions
- Minimal number of different cladding and waterproofing systems to limit termination and transition detailing and trades involved
- Minimal reliance on sealant systems for termination and transition detailing
- Joints designed to shed water
- Minimal reliance of single-barrier systems
- Secondary systems installed where practicable, including
  - Flashing
  - Dampproofing
  - Weeps
  - Drainage tubes
- Proper allowance for thermal expansion, contraction, and weathering cycles
- Absence of level or horizontal envelope areas that would allow ponding water at roofs, balconies, and walkways
- Drainage of water away from an envelope as quickly as possible both above and below grade; gutters, drains, slopes, drainage mats are used where appropriate
- Recessing windows and curtain walls at slab edges
- Adequate space provided to detail all termination and transition details properly
- Preconstruction and envelope review meetings with all trades, manufacturers, and supervision that will affect envelope performance involved
- Testing and review of detailing where necessary to ensure effectiveness before construction begins
- Joint and several warranties for all envelope components
- Quality control procedures during construction

- No substitution of materials or systems after approvals, testings, and reviews
- Proper envelope maintenance

Successful envelopes require reviewing all building envelope components, including how they interrelate. To understand these requirements, start at below-grade construction and work upward, exploring and reviewing the qualities and designs necessary for an effective building envelope with the longest life cycling possible. Refer to the typical envelope building detail shown in Fig. 8.6.

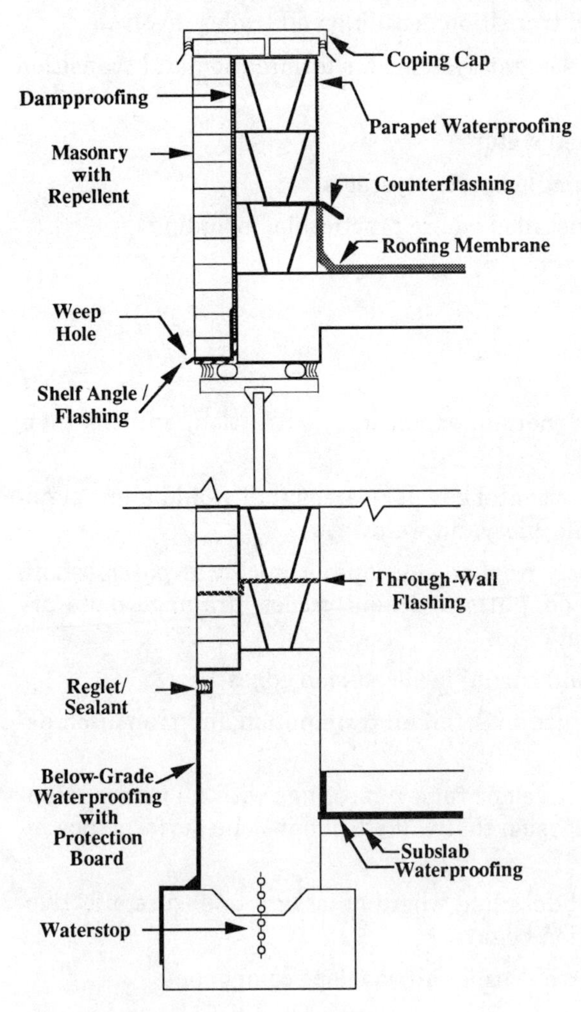

Figure 8.6 Building envelope detailing.

Wall-to-floor junctions are areas susceptible to leakage. These areas should receive waterstops for additional protection with waterproofing materials installed properly along this intersection with necessary cants and reinforcements. Installation details of floor membranes at these locations are continued up to transitions with vertical or wall membranes. If necessary mud slabs should be used to provide a sound substrate on which to apply horizontal waterproofing membranes.

Similar detailing should continue at any below-grade pits or structures such as elevator or escalator pits. Pile caps and similar structural foundations should be wrapped completely and continually and transitioned into floor or horizontal membranes. Structural design may prohibit membranes from being applied between concrete pours such as floor-to-wall details. This requires wrapping membranes beneath foundations and up sides to allow for unbroken continuity.

Transitions from below- to above-grade waterproofing should be watertight, but they also must allow for thermal and differential movement that occurs at this intersection. Reglets, flashings, or other means of protection should be installed to provide for this transition.

The above-grade envelope must then be carried completely up vertical surfaces and tied into the horizontal or roofing portion of the envelope. This horizontal portion is in turn tied back into the opposite side of vertical envelopes, back to below grade, forming a complete envelope on a structure.

All transitions and terminations in a vertical envelope must be completely water and weather resistant. Transitions between such features as walls to window frames must be waterproof and allow for thermal and differential movement. In this case, it is typically a well-designed and installed sealant joint. Sealants are frequently used for transitional waterproofing. However, sealants are often overused when better materials or different systems, such as flashings, should be used.

Weathering of exposed envelope systems creates movement in all materials, and allowances must provide for this movement. Movement above grade is created by several phenomena as summarized in Table 8.2.

To provide for this movement and volume change, control and expansion joints must be designed and placed where such movement is expected. Among the envelope locations for placement of control joints are

- Changes in materials
- Changes in plane
- Material volumetric expansion

TABLE 8.2 Common Envelope Movement Causes

Movement	Cause
Thermal	Expansion or contraction movement caused by temperature changes
Structural	Caused by the curing process of concrete during settlement or structural loading of the building
Differential	Materials have individual coefficients of movement, which will differ from surrounding materials, causing differences in movement between the materials
Moisture content	Certain materials, particularly masonry, swell when subjected to wetting; this movement or enlarging, when calculated as an aggregate total of the entire facade area, can be considerable

- Construction joints
- Junction of facade materials to structural components
- Changes in direction
- Concentration of stresses (such as openings in a structural wall)

All such construction details in a composite wall area should be reviewed, and, where appropriate, control or expansion joints installed. Sealants installed in these areas should be completed according to application requirements presented in Chap. 4.

All appurtenances on a building should be checked for watertight integrity. Often-overlooked items in this category include exhaust ventilators, fresh air louvers, mechanical vents, signage, lightning equipment, pipe bollards, and mechanical and electrical piping. All should be watertight and weather resistant, including transitions into adjacent materials.

Envelope review then proceeds to roof areas. Roofing systems must be adequately transitioned into the adjacent wall system. This is accomplished either with flashings and counterflashings on a parapet or edge flashing directly covering adjacent wall facades. As with vertical portions, roof areas must allow for movement with adequate expansion and control joints. Additionally, all surfaces should be sloped to shed water as quickly as possible.

Parapet walls create a particular problem in life cycling of a structure. Many unique stresses occur at parapet walls due to the imposed designed loads. Parapet walls move both vertically and horizontally due to thermal movement. In addition, both sides of parapet walls are exposed to weathering which exemplifies this movement of horizontal and vertical expansion and contraction.

Introduction of an adjacent roof slab, which expands against or contracts away from a parapet, imposes a great amount of additional stress on a parapet wall. These stresses may cause bowing of parapet walls and cracking of facing materials or systems that lead to water intrusion. Once begun, this entering water imposes additional stresses such as swelling, freeze–thaw cycling, and corrosion of reinforcement. Often parapets eventually fail to function properly.

When unbreathable coatings are applied to a roof side of parapets, such as black asphalts, heat absorption into parapets increases. This type of waterproofing can damage the integrity of a parapet structure. The numerous situations involved with parapet construction require special designs to ensure that sufficient structural components as well as control and expansion joints are included in construction.

As with waterproofing systems, water leakage through roof areas is typically located within 1 percent of the entire surface area, most often occurring at termination and transition details. This often occurs at equipment supports, equipment pads, plumbing stacks, scuppers, drains, skylights, and lightning and electrical equipment. Detailing transitions properly and providing differential movement at these areas ensures watertight transitions to roofing materials. Movement allowance for roofing details includes movement created by vibrations from mechanical equipment.

Roofing Review

There are frequently more subcontractors and different tradespeople involved in a typical roof installation and related construction than in any other envelope component. Trades often involved in roof construction include

- Roofing
- Mechanical and HVAC
- Carpentry
- Masonry
- Miscellaneous metalworking
- Waterproofing
- Concrete deck installing
- Sheet metaling
- Electrical
- Curtain or window wall contracting

Among the related envelope components involved within a typical roofing envelope are

- Roof drains
- Lightning rods
- Balustrades
- Mechanical equipment
- Electrical equipment
- Signs
- Copings
- Skylights
- Parapets
- Penthouse walls

It is important that each of these envelope components be completely waterproof itself to ensure a roofing envelope's effectiveness. Terminations and transitions necessary to incorporate these trades and systems into an envelope are candidates for errors and resulting water infiltration. This multiple discipline requirement highlights the importance of requiring a preconstruction meeting that involves all parties who affect roof performance and resulting envelope watertightness.

This conference must include electricians, mechanical contractors, and curtain wall and waterproofing subcontractors, in addition to any of those people listed above who are included in specific project requirements. Each contractor must be made aware of his or her responsibility to interact with all other trades for successful completion of an envelope.

Terminations and transition details are reviewed and require the manufacturer's preapproval of all project details. This review should include discussion and resolution of the following frequent causes of envelope infiltration related to roofing:

- Inadequate and improper transition and termination details
- Inadequate drainage (no flat roofs) and absence of testing for proper slopes before installation
- Too many separate roof penetrations
- Too much equipment and traffic on roof areas

Manufacturers should be consulted by designers to review proposed detailing to ensure the system will adequately function under the proposed job site conditions. Any unusual conditions expected to be encountered such as equipment penetration and traffic on roof should be carefully reviewed to ensure a material's adequacy.

During project bidding stage, manufacturers should preapprove proposed installers and allow only those contractors who are familiar with the procedure and trained to compete for the roof installation contract. This coordination should continue through the actual installation, with reviews and inspections conducted as necessary by the manufacturer. Finally, by requiring joint manufacturer and contractor warranties, the manufacturer continues its involvement throughout the warranty period. Warranties are discussed in detail in Chap. 9.

1/90 Percent Rule

Throughout this book, emphasis is given to proper selection and installation of envelope waterproofing systems. As this chapter has shown, however, successful installation goes beyond selection and application of a single envelope component. Only if all individual components of a building's envelope have adequate transitions with one another will a building remain watertight and weather tight.

This is especially true of buildings that use a variety of composite finishes for exterior surfacing such as brick, precast, or curtain wall systems. These designs incorporate a variety of waterproofing methods. Although they might each act independently, as a whole, they must act cohesively to prevent water from entering a structure. Sealants, wall flashings, weeps, dampproofing, wall coatings, deck coatings, and the natural weather tightness of architectural finishes themselves must act together to prevent water intrusion.

As much as 90 percent of all water intrusion problems occur within 1 percent of the total building exterior surface area.

Field construction is predicated on bringing numerous crafts together into the completion of a structure. Too often these crafts are supervised and inspected independently of one another without regard for coordinating their work into a solidarity effort.

Quality of field construction must be expanded to monitor and supervise the successful installation of transitions and terminations of envelope components. Project plans and specifications by the architect and engineer must clearly detail the responsibility for this work. Contractors must then take the responsibility for supervising and coordinating proper installations. Building owners must implement maintenance programs required throughout envelope life cycling.

Intrusion of water and weather at any envelope detail will create further problems by compounding itself. Leakage promotes deterioration of substrates and structural reinforcement that begins and accelerates throughout the entire process. Each cause feeds the other: further leakage causes further deterioration, further deterioration causes further leakage, resulting in eventual building envelope failure, damaging the structure and interior contents.

Such action results in the lawsuits, wasted energy, increased repair costs, loss of revenue, and inconvenience to tenants so frequent in the construction industry. Applying the 1/90 percent principle prevents these situations.

Chapter

9

Safety and Maintenance

Owners, contractors, and employees must abide by construction safety regulations that prevent unsanitary, unsafe, and hazardous working conditions that affect an individual's health and safety. Disregard of government regulations may result in punishments ranging from fines to imprisonment. What is more important, improper working conditions can cause death or severe injury to employees and bystanders (Fig. 9.1).

Laws and regulations that govern the use and installation of a waterproofing system must be thoroughly researched. Such regulations are available from government offices that implement policies, including

- Occupational Safety and Health Administration (OSHA)
- Environmental Protection Agency (EPA)
- Department of Transportation (DOT)
- State and local regulations

These agencies have specific and detailed regulations for the use and installation of building envelope materials, general construction practices, and hazardous chemical use. Field inspectors ensure compliance of regulations and are empowered to levy fines, violation notices, and penalties that can lead to prison sentences. For complete protection, contractors completing work must have written and enforced safety policies and hazardous waste programs, ensuring compliance with government regulations.

Figure 9.1 Poor safety procedures can often result in accidents. (*Courtesy of Western Group*)

Occupational Safety and Health Administration

OSHA has set extensive standards for occupational safety and health in all workplaces. Specific OSHA regulations and enforcement policies cover all construction worksites, including renovation of existing structures. No contractors are immune from OSHA guidelines. Regulations for the construction industry cover a broad scope, including

- General safety and health
- Environmental provisions
- Material handling, storage, use, and disposal
- Ladders and scaffolding
- Personal protective equipment
- Fire protection
- Signs, signals, and barricades
- Hand and power tool requirements and usage
- Welding and cutting
- Electrical wiring and equipment
- Floor and wall openings

- Cranes, elevators, and lift hoists
- Motorized vehicles and equipment
- Excavations, trenching, and shoring
- Demolition
- Explosives
- Power transmission
- Rollover and overhead protection

Whereas all these standards may relate to a specific installation, the first four are most important for waterproofing systems and building envelopes. The following data present a general review of these regulations but are not meant as a substitution for a complete review of current regulations.

General safety and health provisions

General safety and health regulations include specific requirements for first aid and safety training of personnel. Minimum standards for first aid equipment, including adequate fresh drinking water and sanitation facilities, must be at job sites and readily accessible. Emergency numbers, such as those for hospitals, must be posted in a conspicuous place. Illumination levels, sound levels, and requirements for protection of chemical gas vapors and dirt are specifically regulated. These provisions also include standards for handling and removal of asbestos.

Personal protection

The OSHA regulation section on personal protection contains specific requirements for personnel safety and storage of fire hazardous materials. Requirements include hard hats, eye and face protection, and respiratory protection. Fire protection and fire fighting equipment must be provided at all projects where hazardous materials or systems are being used.

Specific regulations cover the types of fire extinguishers required and storage requirements for fire-rated materials. These materials are referred to as red label materials because of red warning labels attached. Common waterproofing systems requiring warning labels include sealants, solvents, and deck coatings.

Signs, signals, and barricades

Regulations for signs, signals, and barricades are for the safety of construction personnel and pedestrians near or on a construction site. Signs warning of specific dangers are regulated as to size, lettering, and colors. Barricades are required to deter the passage of unauthorized pedestrians or vehicular traffic into a dangerous area.

Material handling, storage, and disposal regulations

Material handling, storage, and disposal regulations cover requirements during material use and storage at job sites and the proper disposal. Also regulated are lifting and rigging equipment used to transport materials vertically and the means by which materials are placed into appropriate refuse containers. Specific regulations for disposing hazardous materials are governed by the Environmental Protection Agency not OSHA.

Ladders and scaffolding

Ladders and scaffolding is an important part of OSHA standards; it deals with issues causing frequent accidents and accidental deaths. The size, type, construction, and placement of ladders are specifically detailed. No ladder should be permitted at or used on a construction site that is not clearly marked as OSHA approved. Ladders used for access to roofs or other landings must extend 36 in above the landing and be tied securely to structures to prevent slippage or displacement during their use.

Suspended or swing stage scaffoldings, which present a dangerous working condition, should be carefully inspected. Both scaffold construction and its rigging must conform to safety regulations set forth by OSHA. (See Figs. 9.2–9.4.)

Mechanics must wear a safety harness (safety belts are no longer acceptable) securely attached to a structural building component. Attachment must be independent of scaffolding rigging and be tied to building items such as structural columns. Plumbing stack pipes, exhaust fans, and similar items will not withstand the force of a falling person or scaffold and should not be part of the rigging attachment under any circumstances.

This OSHA section also covers the composition and attachment of suspension scaffolding in detail. Actual stage deck construction, including use of toe boards, specification of size and height of back rails, and use of safety netting, are regulated. Scaffolds must be constructed to support a load of at least four times the intended load. Regulations also cover material composition and the size of cable used to support scaffolding and requirements for attachment to a structure. Cable must carry at least six times the expected loading.

Stack or tubular scaffolding requirements include compositions, height limitations, planking, and anchoring. Scaffoldings are set on

Figure 9.2 Proper low-rise rigging and safety procedures. (*Courtesy of Sto*)

foundations rated to support the maximum loading to be encountered. Scaffolds must be secured at a minimum 30 ft horizontally and 26 ft vertically. Any scaffolding constructed and used that is more than 125 ft high must be designed and approved by a registered engineer.

Figure 9.3 High-rise rigging requiring use of building's structural components.

Figure 9.4 High-rise scaffolding supported by building's structural components.

These are just a few of OSHA's regulations. Anyone endeavoring to complete construction, no matter how minor in size, must abide by all OSHA regulations.

Department of Transportation

The federal Department of Transportation has greatly expanded its regulations affecting transportation and delivery of hazardous materials more than 1000 lb per shipment. This is roughly equivalent to two drums of materials, or 100 gal. Since many waterproofing materials are considered hazardous materials, DOT regulations affect their transportation and delivery. Federal auditors, state and city police, and weight station personnel enforce these regulations. Drivers of transport vehicles, as well as company management personnel, can be found in nonconformance of regulations and can be penalized or imprisoned.

DOT regulations include requirements for both the vehicle in which the materials are transported and the driver transporting materials. Vehicle requirements include vehicle registration and assigning of registration numbers that must be conspicuously posted on both sides of a vehicle.

The purpose of registration is to ensure that vehicles meet DOT guidelines and to provide contact with companies for monitoring by DOT. Vehicles transporting hazardous materials must display appropriate warning placards (e.g., flammable, combustible, and radioactive). These signs are posted only when a vehicle contains such materials; otherwise placards must be covered or removed.

Regulations also require that a vehicle file be maintained. This file includes identification numbers, maintenance records, and pre- and postinspection records that indicate that any deficiencies found were corrected. Proper shipping records must be available to drivers while transporting materials. All materials must be shipped by DOT approved methods and in DOT approved packaging. Vehicles transporting hazardous materials must have adequate safety provisions such as fire extinguishers.

DOT also has strict regulations governing drivers of vehicles transporting hazardous materials. Drivers must have a commercial driver's license, a record of a physical examination every 24 months, and certification of having passed a written examination of *Federal Motor Carrier Safety Regulations.* They must also pass an employment check, including past employers and motor vehicle records check for accident and traffic violations.

Most waterproofing materials are governed by DOT hazardous, flammable, or dangerous categories. This requires a thorough knowledge of applicable laws and required record keeping. Waterproofing contractors should contact the Department of Transportation to receive appropriate information relating to transportation of hazardous materials.

State and Local Agencies

In addition to federal regulations, most state and local governments have regulations and enforcement policies unique to their specific area. For instance, state Department of Transportation agencies may add to the federal requirements concerning transporting hazardous systems and other building envelope components.

Local codes enforce hazardous waste collection and related storage sites. Most states require registration for hazardous waste disposal. OSHA regulations are also enforced locally by state and local governments, and local agencies should be contacted for specific regional regulations.

Ignorance of the law is not an excuse for avoiding requirements affecting the use, installation, transportation, and disposal of hazardous, flammable, or dangerous materials. Efforts must be made to be acquainted completely with these laws and regulations in order to avoid possible penalties and even imprisonment.

Material Safety Data Sheets

The right to know law protects employees from exposure to dangerous or hazardous chemicals. This law requires material safety data sheets to be published by every manufacturer of such materials and to be provided to all employers for distribution to employees. These data sheets must be readily available at every project for review by employees to educate themselves on dangers of the material's use and risks to health. Specific safety and handling requirements to ensure the safe and proper material usage are also included.

These safety data sheets have a specific written form (OSHA Form 20) and include the following information:

- Manufacturer's name and address
- Chemical name and family
- Any hazardous ingredients in the material
- Physical properties
- Fire and explosive data
- Health hazard data
- Reactivity data
- Spillage or leak procedures
- Protection information
- Special precautions

All safety precautions listed on sheets must be complied with. OSHA inspectors will not only inspect usage compliance but also ensure that data sheets are available to all employees.

Should an accident occur involving a hazardous material, data sheets are given to doctors treating the patient. This allows review of the material's chemical composition and of appropriate treatment options. Additionally, should a material spill or leak at a job site, data sheets provide steps to be taken and means for proper disposal.

Environmental Protection Agency

Any organization that generates hazardous waste must register with the U.S. Environmental Protection Agency and receive an EPA identification number. The EPA issues regulations that monitor the use, storage, transfer, shipping, or disposal of hazardous waste. This agency conducts on-site inspections and is authorized to enforce and prosecute firms or employees that do not comply with such regulations. Recently, OSHA has agreed to joint project inspections of job sites with EPA inspectors. These regulations are part of the Resource Conservation and Recovery Act (RCRA). Specific regulations apply for waste generators based on the amount of waste the firm generates on a monthly basis.

Conditionally exempt are firms that generate less than 25 gal of hazardous waste. Small-quantity firms produce 25–300 gal per month; generator firms produce more than 300 gal per month.

Since waterproofing systems often fall under the classification of hazardous materials, they are governed by the RCRA. Solvents, thinners, and primers used in conjunction with waterproofing are also regulated. Examples of hazardous waterproofing materials include coatings containing solvents such as solvent-based urethanes and most epoxies. Materials become hazardous waste when any of the following situations occurs:

- Unused materials remain in their original container.
- Materials reach their shelf life (usable time).
- Spillage of the material occurs.
- Hazardous materials are introduced into nonhazardous systems.

A proper identification number is obtained from the federal agency by completing EPA form *Notification of Hazardous Waste Activity.* State and local government agency registrations must also be complied with.

Once all necessary licenses are obtained, proper storage and periodic disposal methods are implemented. EPA, state, and local governments regulate the amount and means of hazardous waste storage. Only recognized authorized and licensed companies should be used to dispose of accumulated waste materials. EPA regional offices or local governments provide listings of such firms.

Building owners should be aware of these regulations and ensure that contractors who are employed to complete work on their building or structures are in compliance with them. Additionally, owners should not permit contractors to dump waste materials into building waste receptacles or trash dumpsters. This action may result in fines and penalties levied against the building owner.

Contractors

Regardless of the quality, performance characteristics, and cost of a waterproofing system, the systems are only as effective as the caliber of installation. Even the best systems may prove worthless or ineffective if not installed and transitioned properly into other envelope systems.

Considering that most construction systems are field manufactured, it is mandatory for properly trained mechanics and competent contractors to complete installation of any envelope component. Ineffective installation cannot only destroy the performance of a material itself, but it can also lead to structural and interior contents damage. This results in costly repairs and loss of revenue for a building owner.

By prequalifying and selectively choosing competent and experienced contractors, unnecessary problems are eliminated during life cycling. Additionally, this process should ensure that the contractor will be available for repairs and will honor warranty items that may occur. There are many qualities a potential contractor should possess, including

- Experience in the specific type of installation
- Properly trained mechanics to complete work
- Certification by material manufacturer
- Organized and enforced safety policy
- Payment and performance bonds for total contract sum
- Insurance as required by federal, state, and local laws
- Joint manufacturer and contractor warranties
- Availability of maintenance bonds for warranties
- Sufficient equipment to complete installation
- Financial and customer references
- In-place hazardous waste programs
- Structural quality programs

When selecting a contractor, all of the above points should be considered. Reliance solely upon low bids often ends up costing more in maintenance and repairs over life cycling of installed systems.

Bonding capacity is a reliable means of allowing only responsible firms to complete work. Bonding and insurance companies run extensive background investigations of contractors before providing either bonds or insurance to a firm. Upon investigation of the contractor's experience, financial, and other capacities, bonding companies will set bonding limits in dollar amounts for the contractor.

Bonds act as insurance policies for the benefit of an owner. Requiring contractors to provide a payment and performance bond in the full contract amount assures owners that the contract will be completed and all materials, suppliers, and employees will be paid. If not, a bonding company will take over the contract and complete the work and insure payments of all outstanding invoices.

Likewise, maintenance bonds ensure that warranties will be honored even in the event a contractor goes out of business before the warranty expires. All bond premiums will be added to the cost of a project, but they offer protection that otherwise may not be available for performance and warrantability.

Field mechanics installing waterproofing systems have ultimate control over success or failure of the in-place waterproofing system. Field mechanics must be properly trained and motivated to complete all installations in a professional manner. Supervision must be provided to ensure that installations act cohesively with other envelope components.

Although owners can require that a mechanic have experience and training to install materials, contractors have ultimate control over job site and working conditions. These conditions include wages, benefits, and safety conditions, all of which influence installation quality.

Finally, warranties provided by a contractor should cover both the labor and materials. Questions to consider in reviewing a proposed warranty include

- Will it cover material failure?
- Whose responsibility is it to uncover buried or covered systems?
- Are consequential damages (interior contents) covered?
- Is the warranty bondable?
- Is there a dollar limit to the repairs?
- What escape clause does the warranty contain (e.g., structural settlement)?
- Is the warranty issued jointly by the manufacturer and contractor?
- Does the warranty cover transition and termination detailing?

If necessary, proposed warranties should be reviewed by counsel and necessary changes should be made to protect the owner before a contractor is awarded the contract.

Manufacturers

Equally important to the success of a waterproof system is the experience, assistance, and reputation of the material manufacturer. The quality of the material manufacturing process itself is of high importance, but choosing a manufacturer should take in considerations beyond quality, including

- Manufacturer warranties
- Length of time product has been manufactured
- Sufficient number of previous installations
- Adequate testing and test information of product
- Detailed installation instructions
- Availability of representatives to review installations

- Certification and training of applicators
- Maintenance instructions
- Material safety data sheets
- Manufacturer's assistance in specification preparation and detailing termination and transitions

The manufacturer's representations for a particular product should be reviewed and test results compared to similar products. Test results based on recognized testing laboratories such as ASTM should be consulted. This testing allows materials to be compared with those of other manufacturers as well as completely different systems.

Manufacturer warranties should also be carefully reviewed. Often a warranty only covers material failure and provides no relief for building owners in case of leakage. Considering that all waterproof systems require field application or construction, too often it is easy for a manufacturer to pass liability on, citing poor installation procedures.

Therefore, warranties should cover both labor and materials. This places requirements on a manufacturer to ensure that only experienced mechanics and contractors install their products. A labor and materials warranty from a manufacturer and contractor provides protection when one or the other goes out of business. It also prevents attempts to place blame elsewhere when there is a question of liability for repair problems.

Maintenance

No building or structure is maintenance free. In fact, of total costs, 30 percent consist of original construction costs and 70 percent of maintenance costs. Considering the possible damage and costs that might occur, it is just as important to maintain the exterior as the interior of a building. Regular exterior maintenance prevents water intrusion and structural damage that might be associated with water infiltration.

An effective maintenance program involving the building envelope depends on using qualified inspection procedures to determine the required maintenance. A building requires complete inspection from top to bottom, including a review of all exterior elements, at recommended intervals of every 5 years but no longer than every 10 years.

Any building portion inaccessible by ordinary means may require hiring a contractor for scaffolding and inspection. Only competent building trades personnel should make these inspections, be it an architect, engineer, or building contractor.

In view of 90 percent of all leakage being caused by 1 percent of the building envelope, all components of an envelope must be inspected. All details of inspection, including exact locations of damage and wear, that will require attention after an inspection should be documented.

Among envelope components, the following require complete and thorough inspection:

- *Roofing,* with particular attention to terminations, flashings, protrusion, pitch pans, skylights, and copings
- *Above-grade walls,* with attention to expansion and control joints, window perimeters, shelf angles, flashings, weeps, and evidence of pollutant or chemical rain deterioration
- *Below-grade walls,* checking for proper drainage of groundwater, signs of structural damage, and concrete spalling
- *Decks,* with attention to expansion and control joints, wall-to-floor joints, handrails, and other protrusions

These are only the highlights of maintenance relating to waterproofing materials. Inspection procedures for existing damage and buildings that have not been maintained are discussed in Chap. 7.

During inspections, effectiveness of a waterproof system should be monitored. This includes water testing if necessary to check systems already in place. This requires inspection for items such as clogged or damaged weeps, cracks or disbonding of the elastomeric coating, deteriorated sealants, damage to flashings, and wear of deck coatings.

Most waterproof systems require maintenance procedures of some type, and these recommendations should be received from the manufacturer. Certain items will require more maintenance than others, and provisions need to be made to monitor these systems more frequently. For instance, vehicular traffic deck coatings receive large amounts of wear and require yearly inspections. Maintenance for traffic areas includes replacement of top coatings at regular intervals to prevent damage to base-coat waterproofing.

Dampproofing behind a brick veneer wall usually requires inspection to ensure that the weeps continue to function. Other unexposed materials, such as planter waterproofing systems, may require more attention; they should be checked to ensure that water drainage is effective and protection surfacing remains in place during replantings.

Exposures to elements affect waterproofing systems, and required maintenance often depends upon this exposure. Factors affecting waterproofing systems include

- Thermal movement
- Differential movement

- Weathering
  - Ultraviolet exposure
  - Freeze–thaw cycles
  - Rain
- Wind loading
- Chemical attack
  - Chloride (road and airborne salt)
  - Sulfides (acid rain)
- Settlement
  - Structural
  - Nearby construction
  - Acts of God (hurricanes, earthquakes, tornadoes)

Regular maintenance inspections should monitor any damage that might be caused by these types of wear and weathering, and repairs should be completed promptly according to the manufacturer's recommendations. Manufacturers should make representatives available to assist in the inspection and to make recommendations to the building owner if repairs or maintenance work is necessary.

Should a particular portion of an envelope be under warranty by a manufacturer, contractor, or both, necessary maintenance or repairs should be completed by firms warranting these areas. This prevents nullifying warranties or obligations of a manufacturer or contractor by allowing others to complete the repairs. If extended warranties are available, manufacturers should be consulted for proper maintenance procedures. For example, a contractor or manufacturer can provide a 5-year warranty plus an optional 5-year renewal. This requires that after the initial 5-year period, manufacturer and contractor make a complete inspection, at which time all necessary repairs are documented. Upon an owner's authorization for repair completion and payment for these repairs, the manufacturer or contractor extends the warranty for an additional 5 years.

Warranties

It would require several legal courses to cover warranties and guaranties and their legal consequences completely. This section therefore approaches the subject as summarized in the phrase "Let the buyer beware." Due to all intangibles involved in field construction, it is rare to find any warranty that simply states "System is guaranteed to be waterproof." All warranties typically exclude circumstances beyond the manufacturer's or contractor's control. However, owners should review warranties to ensure they are not full of exclusion clauses that negate every possible failure of material or installation, making the warranty in effect worthless.

No warranty is better than the firm that provides it, and should the manufacturer or contractor go out of business, a warranty is useless unless bonded by a licensed bonding company. A manufacturer or contractor that places emphasis on its reputation may be more likely to take care of repairs or warranty items regardless of the limitations that appear in the warranty or guarantee.

The terms *warranty* and *guaranty* are typically used interchangeably and have no distinct difference. In preferential order, following is a list of warranty and guaranty types:

- Bonded
- Joint manufacturer and contractor warranty
- Combination of separate contractor and manufacturer warranties
- Manufacturer warranty covering both installation and materials
- Contractor labor warranty only
- Manufacturer material warranty only

Any of these warranties must be specific to be enforceable. Clauses such as "warrant against leakage" leave open the responsibility of a contractor or manufacturer. Does this mean any leakage into the building or leakages only through the applied systems? What happens if a juncture between the warranted system and another system fails? Who covers this failure? Regardless of the warranty type, it should be specific as to what is and is not covered under guarantee terms.

Bonding of a warranty provides complete protection for a building owner and insures against failure by both contractor and manufacturer. A bonded warranty should be underwritten by a reputable, rated bonding company licensed to do business in the state in which it is issued.

Some bonded warranties may limit the extent of monies collectible under warranty work. This works as a disadvantage to an owner should a system require complete replacement. For example, consider an inaccessible system, such as below-grade waterproofing, where costs for obtaining repair access can well exceed the actual cost of repairing the leakage.

Joint warranties, signed by both contractor and manufacturer, offer excellent protection. This warranty makes both firms liable, jointly and severally. This ensures that if one firm is not available, the other is required to complete repairs. These warranties typically have separate agreements by manufacturer and contractor, agreeing to hold each other harmless if repairs are clearly due to defective material or defective workmanship. This separate agreement does not affect the owner, as the document issued makes no mention of this side agreement.

Manufacturers are selective with whom they choose to become signatory to such agreements. They qualify contractors financially and provide training and make available manufacturer's representatives to ensure materials are installed properly. Additionally, most manufacturers thoroughly inspect each project installation and require completion of its own punch list before issuing warranties.

Warranties that cover labor and materials separately have consequences an owner should be aware of. By supplying separate warranties, contractors and manufacturers often attempt to pass blame to each other rather than correct the problems. Owners may have to contract out work to other parties to complete repairs and attempt to recover from the original manufacturer and contractor by legal means. In some situations, this method is used to provide warranties of separate length for labor and material (e.g., 5-year labor, 10-year material). These have the same limitations and should be reviewed carefully to combine the two into one agreement.

Other warranties—labor only or materials only—have limited protection and should be judged accordingly. Since most systems are field installed, labor is most critical. However, materials can fail for many reasons, including being used under wrong conditions. Therefore, both materials and labor should always be warrantied. By not requiring a material warranty, manufacturers may not be under obligation to ensure that materials are being used for appropriate conditions and with recommended installation procedures.

Actual terms and conditions of warranties vary widely, and assistance from legal counsel may be necessary. For common warranty clauses, special attention should be made to the following:

- Maintenance work required of an owner to keep the warranty in effect
- Alterations to existing waterproofing systems that can void a warranty
- Wear on systems that may void the warranty (e.g., snowplows, road salting)
- Unacceptable weathering (e.g., hurricanes, tornadoes)
- Owner's rights to repair or replace the system limited by
  - Replacement of materials only
  - Specific dollar amount

Requiring access for repairs (such as excavation of below-grade areas)

Limitations on consequential damages

- Requirement that prompt notification is given, usually in writing, within a specific time
- Contractor and manufacturer refunding of original cost to satisfy warranty instead of repairs
- Specific exclusions of responsibility

Structural settlement

Improper application

Damage caused by others

Improper surface preparation

- Complete replacement of a faulty system versus patching existing system

All warranties are limited and must be reviewed on an individual basis to eliminate any unacceptable clauses before signing the contract, purchasing materials, and installation. Items such as the actual specific location of a waterproof system should be clearly included in the warranty and not limited to the building address. The terms of what is actually covered should also be addressed (e.g., installation, leakage, materials, or all three). The warranty should be specific, allowing those interpreting a warranty years later to understand the original intended scope.

After project completion, warranties are typically an owner's only recourse and protection against faulty work and materials. With this in mind, warranties should be given the same close scrutiny and review as the original design and installation procedures to protect the owner's best interests.

Finally, recalling the 1/90 percent rule, all too often transitions and terminations are not specifically included in each of the envelope component warranties. By making contractors and manufacturers responsible for the 1 percent of a building's area that create 90 percent of leakage problems, their attention is directed to this most important waterproofing principle. By including these areas in warranties, contractors and manufacturers are prompted to act and ensure that these details are properly designed and installed. This prevents numerous problems during life cycling of a building or structure.

Chapter

10

Envelope Testing

There are several steps, methods, and means to test individual or complete portions of a composite envelope. These tests begin with the manufacturer's testing, which ensures that materials are suitable for specified use, longevity, and weathering. Next, an entire composite envelope sampling is tested to ensure that all components, when assembled, will function cohesively to prevent water infiltration.

No project is built or renovated without some testing having been completed. Too frequently, however, the only testing completed, that of material systems by manufacturers, is insufficient to prevent problems that continue to occur at the job site.

Rarely is attention given to testing the 1 percent of a building envelope that creates 90 percent of the water intrusion problems. This 1 percent of a building's area, the terminations and transitions of various independent systems, never is fully incorporated into proper testing.

When Testing Is Required

Testing frequently is used to test new designs, materials, or systems. However, envelope designs that incorporate standard materials also require testing under certain circumstances. For example, masonry walls constructed of typical brick composition but having intricate detailed slopes, shapes, and changes in plane should be tested. Testing in these cases will determine whether flashings as detailed will perform adequately in the various detail changes incorporated into the design.

Testing should also be completed when envelope components contain areas such as expansion joints in unusual or previously untested areas; for example, a sealant expansion joint in a sloped area that may pond water.

Specially manufactured products, such as specially colored sealants, brick manufactured in unusual textures, metal extruded in unusual shapes, and joints are examples of envelope components that should be tested to prevent problems after complete envelope installation.

Any time a new design comprising several different materials is developed for a proposed envelope mock-up, testing is imperative. This is particularly true for high-rise construction.

Cladding materials used in today's designs and construction are lighter weight and thinner, requiring less structural materials and supports. This lowers overall building costs but in turn presents numerous problems in envelope effectiveness, particularly in watertightness. This is in comparison to the massive masonry walls of more than 1 ft thickness used in early high-rise construction, where shear magnitude of the envelope eliminated the need for such testing.

Testing Problems

Manufacturers, although making recommendations for termination or transition detailing, will not incorporate these areas in their material testing. A manufacturer of sheet-good membranes will test the actual rubber materials for weathering, elastomeric capabilities, and similar properties. The manufacturer will not, however, test the adhesive material used to adhere materials to a termination detail for weathering, movement characteristics, and so forth.

Likewise, transition details, such as between above-grade and below-grade areas, detailed by an architect are not tested by either the waterproofing or dampproofing material manufacturers. Lack of testing in these and similar details of the 1 percent of the building's area reveals another reason for the continuing cause of these areas contributing 90 percent of envelope water infiltration problems.

Mock-up testing of a building envelope portion often eliminates replication of termination and transition details. Testing often is completed on envelope curtain wall portions only and does not include masonry portions, transitions from curtain wall to parapet wall, coping, and roofing transitions.

Ensuring that each material or system is tested independently does not ensure that the composite envelope when completed will be successful. Any material, even if it performs singularly, does not ensure that the composite envelope will be successful.

These are some reasons testing fails in preventing water infiltration. As long as buildings are manufactured from a variety of systems, which must be transitioned or terminated into other components and these details are untested and improperly installed, problems will continue to occur.

Standardized Testing

A thorough review of all testing available including manufacturer's testing, independent testing, laboratory testing, and site testing should be made. This will make sure that the entire envelope is properly tested to ensure watertightness and envelope longevity.

Available testing includes

- Laboratory analysis
- Mock-up testing under simulated site conditions
- Job-site testing
- Long-term weathering testing

Such tests are completed by both government (including state and local municipalities) and private agencies. The most commonly referred to private agency testing is the American Society for Testing and Materials.

ASTM testing

ASTM was established in 1898 as an organization for establishing standards in characteristics and performance of materials, including some waterproofing materials, particularly sealants and caulking. These standards are used as a basis of comparison among various products and similar products of different manufacturers.

ASTM standards are specific test requirements described in detail to ensure that individual materials are tested in a uniform manner. This allows for a standard of comparison between different manufacturers or materials. Their characteristics can then be compared for judgments of suitability in such issues as weathering, performance suitability, and maximum installation conditions under which a material will function.

ASTM conducts tests on materials furnished to them by manufacturers, or manufacturers document that their materials have been tested according to the ASTM standards. Tests are completed in controlled laboratory conditions. Tests are designed to test a material's maximum capabilities and limits (e.g., elastomeric sealant expansion limits). Also available are accelerated weathering tests to determine if materials are suitable for use in the extremes of weather including freeze–thaw cycling and ultraviolet weathering. These laboratory tests are typically applied to specific materials themselves (e.g., sealants, coatings) but not to the composite envelope systems, transitions, or termination detailing. The frequently referred to ASTM testing adaptable to waterproofing materials are summarized in Table 10.1.

TABLE 10.1 ASTM Testing for Waterproofing Products

ASTM Test number	Test type
D-412	Tensile strength
D-412	Elongation
E-96	Moisture vapor transmission
D-822	Weathering resistance
C-501	Abrasion resistance
E-154	Puncture resistance
D-71	Specific gravity
D-93	Flash point
D-2240	Shore A hardness
E-42	Accelerated aging
E-119	Fire endurance
D-1149	Ozone resistance and weathering
C-67	Water repellency
E-514	Water permanence of masonry
C-109	Compressive strength
C-348	Flexural strength
D-903	Adhesion

Other testing agencies

Laboratory analysis is also completed by the National Bureau of Standards, a federal government agency. The most commonly referred to federal specifications for waterproofing materials are TT-S-227 for two-component sealant performance, TT-S-00230C for one-component sealants, and TT-S-001543 for silicone sealants. Underwriter's Laboratory tests (UL) for fire endurance of specific materials include test UL-263.

Other independent laboratory analysis and testing include firms or organizations established to test specific application or installation uses. These include the National Cooperative Highway Research Program (NCHRP), which tests clear deck sealers used in concrete construction in highway and bridge work.

Local government agencies, such as Dade County Building and Zoning Department's Product Approval Group, are often established to test and approve materials for use in local construction and renovation. This Miami area public agency tests materials to ensure that products will perform in the harsh environment of south Florida. Tests include ultraviolet and hurricane weathering. Such agencies test individually manufactured materials in laboratory conditions to ensure the adequacy of the material alone. They do not perform tests for complete envelope systems but only for individual components.

This testing allows selection of individual materials that will function under an expected set of conditions, including weathering and wear. For example, a deck coating material is chosen that will function under extreme ultraviolet weathering and heavy vehicle traffic. Testing does not, however, determine the acceptability of transitions used between the deck coating and deck expansion joints or whether coatings will be compatible with curing agents used during concrete placement. The tests allow proper selection of individual materials for use in a composite envelope but do not test individual systems joined together in the envelope construction. Several private laboratories are available to complete testing of composite envelopes.

Mock-Up Testing

Independent laboratories are often used to test mock-ups or composite envelope systems. These tests assume that individual components have been tested and will suffice for job-site conditions including extended use and weathering. Independent laboratories test the effectiveness of composite materials against water and wind infiltration. They can create conditions that simulate expected weathering extremes at the actual building site.

These tests are limited, however, in that they do not re-create long-term weathering cycles and temperature extremes and are rarely applied to the entire composite envelope. Testing envelopes after weathering and movement cycles, particularly at the transitions and terminations, is mandatory to ascertain the effectiveness of the details.

An envelope or curtain wall mock-up is constructed at the laboratory site using specified exterior envelope components. They are applied to a structural steel framework provided by the laboratory or framing is constructed specifically for testing. This framework should include flashing and appropriate transition details if they occur in the selected area to be tested. Testing is completed on a minimum floor-to-ceiling segment height of the envelope.

Preferably, testing goes well beyond this minimum to allow testing of the most advantageous and economically feasible portion of the entire envelope. Typically, testing size is 20–25 ft wide to 30–40 ft high.

Particular attention should be given to transition areas between glass and metal to stone, masonry, or concrete areas, parapet areas, and horizontal-to-vertical and other changes in plane, including building corners. Testing of these transitional details ensures their effectiveness against water infiltration. Unfortunately, termination details such as above-grade areas to below-grade areas are not usually feasible for testing purposes. In addition, structural steel supports used in the testing mock-up often cause different test results if the envelope is to be applied to a more rigid frame such as a structural concrete framework.

Testing is also limited to air infiltration and water infiltration. Tests do not include weathering analysis that often deters envelope component effectiveness during movement cycles such as thermal expansion and contraction.

Thinner cladding materials used today are subject to stress by thermal movement and wind loading. Transition and termination details are different based on the thickness of material and movement stress that is expected with the in-place envelope. This makes testing mandatory. It is less costly to correct problems that appear in design and to construct a mock-up than to repair or replace an entire envelope after it is completed and tested by natural forces and weathering.

Testing will also reveal problems that might occur with coordinating the different trades involved in a singular envelope design. For example, how, when, and who installs the through-wall flashing that runs continuously in a masonry, precast concrete, and window wall? Is the same flashing application detailing applicable in all these instances, and so forth?

Situations that arise during testing make it extremely important that mechanics installing a curtain wall at the job site be the same mechanics who install the mock-up construction for testing. Then, should problems arise in testing, they can be resolved with the knowledge carried to the job site. In the same manner it is important that a contractor supervisor be present and participate in mock-up construction and testing to ensure continuity and quality of envelope job-site construction.

Mock-ups in effect become a partnering or teaming concept, with all partners—architect, owner, contractor, and subcontractors—involved. These partners work together to complete mock-up testing successfully and resolve any conflicts or problems before they occur at the job site.

One serious flaw that frequently occurs should not be allowed—using sealant materials to dam up leaks as they occur during testing. Often discovered leakage is simply taken care of by applying sealant. This happens in areas such as perimeters of windows, joints in metal framing, and transition details. Allowing sealant application during testing goes directly against the actual purpose of testing.

If leakage occurs and the sealant was not part of original detailing, it should not be installed until determination of the leakage is resolved. Sealants are not long life-cycle products compared to other envelope components, and they require much more maintenance than typical envelope components of glass, metal masonry concrete, or stone.

Sealants can offer short-term solutions during testing but often do not function during weathering cycles such as thermal movement of the composite envelope. Even if sealants will perform in this detailing, if they are not part of the original design, the owner's maintenance requirements are increased for envelopes that include sealants and their short-life cycles. Such attempts at quick fixes should not be permitted considering the effect and costs committed for testing, as well as the possible long-term effects of using sealants as a stopgap measure.

ASTM provides evaluation criteria for the selection of independent laboratory testing agencies. These tests include ASTM E-669-79, which provides criteria for evaluation of laboratories involved in testing and evaluating components of a building envelope, and ASTM E-548-84, which includes generic criteria for evaluating testing and inspection firms.

Mock-up testing of envelope involves three types of tests, including

- Air infiltration and exfiltration
- Static water pressure
- Dynamic water pressure

Air infiltration and exfiltration testing

The air infiltration and exfiltration test determines envelope areas that will allow air to pass into or out of a structure. Although this test is typically for control of environmental conditions, if air can pass through an envelope, water can also pass through.

Wind loading can force water into a structure, or unequal air pressures between exterior and interior areas can actually suck water into a building. Therefore, a building must be completely weatherproof to be completely waterproof. A completely waterproof building is therefore completely weatherproof.

Air infiltration or exfiltration tests are typically conducted according to ASTM-283. This test is used for measuring and determining any airflow through exterior curtain walls. This is a positive pressure test, meaning that positive pressures are applied to an envelope face.

To conduct this test, a sealed chamber is constructed to enclose the back of a composite envelope portion being tested completely. Air pressures can be lowered in the chamber by removing air and creating a vacuum in the chamber.

The lower air pressure then draws air through the envelope from higher pressure exterior areas. Air is drawn through any envelope deficiencies. This air penetration can be determined by measuring pressure differentials within the chamber. However, if air penetration is occurring, it is difficult to locate specific failure areas during testing.

This testing can be reversed by forcing additional air into the chamber to create higher chamber air pressures. This type of test creates negative pressure on an envelope face. (The explanation of negative and positive air pressure testing is similar to the explanation of negative and positive waterproofing systems.)

With this testing, air that is forced into a chamber can be mixed with fabricated smoke or colored dyes. This allows leakage areas to be easily identified when the colored air begins escaping through envelope components to the exterior. Areas of leakage can then be marked and later inspected for causes. Proper repairs can be completed and areas retested until weathertight.

When testing masonry mock-up panels or curtain walls that contain masonry portions with weeps, it is expected that a certain amount of air will penetrate the envelope through these weeps. The amount of air infiltration that is within satisfactory limits must be determined, and testing must be done to check that infiltration does not exceed these limits.

Static pressure water testing

Static pressure testing uses the same apparatus and methods used with air testing, while at the same time applying water to the envelope face at a uniform and constant rate. Test standards typically used are according to ASTM-E331. Water is applied using spray nozzles equally spaced to provide uniform water application to all envelope components.

By creating positive air pressure on the envelope (withdrawing air from inside the test chamber), water is sucked through an envelope at any point of failure. Areas of leakage are then marked for review and cause determination after testing is completed.

Certain areas, such as weeps built into masonry walls or curtain wall framing, are subject to some water infiltration. As weeps are an integral part of an envelope design, allowable percentages of water infiltration through these areas must be determined. Test results can then be measured to determine the acceptability of infiltration through these areas.

Dynamic pressure water testing

Dynamic pressure testing applies water and wind conditions directly to an envelope face. Airplane prop engines are used to force water against the composite envelope from spray nozzles equally spaced and mounted to frames. This test usually simulates the most severe conditions, such as hurricane and tornado force winds and rain conditions.

Water applied in this manner is forced both vertically and laterally along the envelope face to re-create conditions encountered in high-rise construction. This method of wind loading with the addition of water can force water into envelopes not capable of withstanding designed or expected wind loads.

Glass can bend or flex away from mullions or glazing joints allowing sufficient space for water to penetrate into interior areas. Structural conditions can also change during wind loading to create gaps or voids or even failure of envelope components, allowing direct water infiltration.

Amounts of water and wind introduced onto the envelope can be variably applied to simulate maximum conditions expected at a particular job site. A combination of weather conditions as severe as hurricane forces of 70 mph plus winds and water at a rate of 8–10 in/hour are achievable. This is often the ultimate test for any envelope.

Mock-up testing summary

All three mock-up tests offer excellent previewing of a proposed envelope design. Cost permitting, all three should be completed to determine areas of potential problems accurately. As previously discussed, areas of failure should not merely be sealed using sealants that are not part of the original design. This defeats the purpose of testing, and problems will recur in field construction when envelope watertightness is dependent on improperly designed and applied sealant material.

When envelopes containing masonry walls that combine dampproofing, flashing, and weep systems are tested, water will undoubtedly enter the envelope as designed. The water entering must not exceed the capability of the backup systems to redirect all entering water to the exterior.

Note that laboratory mock-ups can also be used for color and texture approval, limiting mock-ups required at job sites and lowering overall costs of laboratory testing. (See Table 10.2.)

TABLE 10.2 Mock-Up Testing Advantages and Disadvantages

Advantages	Disadvantages
Allows testing of designs and review of problems before actual construction	May not incorporate sufficient termination and transition details
Involves all project participants in reviewing design and offering suggestions for review	Often does not simulate exact job-site structural conditions
Can create conditions beyond the worst expected at actual project site	Does not account for testing after thermal movement and structural settlement changes in actual envelope construction

Job-Site Testing

Tests done at a job site can be as simple or as scientific as required by immediate concerns. Simple water testing using a garden hose and water source is probably the most frequently used means of job-site testing on both new and existing envelopes. Static and dynamic pressure testing can also be accomplished at project sites by laboratories or consulting firms that can provide portable equipment to complete these types of tests.

Test chambers are built at job sites directly over a sample wall portion including curtain and precast wall units. Portions being tested should include as many termination and transition details as possible including any changes in plane. Such job-site tests are limited by actual areas tested but offer the advantage of testing under actual conditions as compared to laboratory mock-ups.

Mock-up panels are often required at job sites to check for color and texture approval before acceptance by the architect. With only a few more construction requirements a mock-up can often be made into fully testable units at the job site.

Mock-ups, besides allowing for watertightness testing, can be used at the site for instructional purposes. This allows an initial means of interaction by all subcontractors involved to make them aware of their role in the overall effectiveness of a watertight envelope. This is especially useful in areas where many subcontractors are involved, such as a building parapet and coping.

The subcontractors are able to work together to develop the working schedules, patterns, and quality required to ensure envelope success before actual installation. Such a process can become an actual partnering exercise in any team building, total quality management (TQM), or partnering program undertaken by an owner and contractor.

There are too many positive benefits that can be derived from testing envelope components at a job site to justify not testing. At minimum, testing should be done immediately after completion of the first building envelope portion during construction. This testing can often reveal deficiencies that can be corrected and eliminated in the remaining areas of construction. Testing can also reveal potential areas of cost savings, better materials, or details that can be incorporated into the remaining envelope portions. (See Table 10.3.)

Manufacturer Testing

Often, manufacturers are depended upon to provide all the testing and information considered for inclusion of their product into a com-

TABLE 10.3 Job-Site Testing Advantages and Disadvantages

Advantages	Disadvantages
Testing can be done on actual construction conditions at a project site	Extreme testing conditions of testing with mock-ups are not possible at job sites
More details of terminations and transitions can be included in actual testing	Larger envelope portions are difficult to test accurately at the job site
Costs are lower since mock-ups are not necessary to construct	Tests are often completed after construction when problems occur

posite envelope. Without proper testing this can lead to numerous problems.

Manufacturers are concerned solely with the materials or system they manufacture. They do not provide necessary information to evaluate their products' usefulness in a composite envelope properly. Their material is not checked or tested for compatibility with adjacent materials used in proposed envelopes.

The specified termination and transition details are often not those tested or typically used by a product manufacturer. A manufacturer often provides insufficient instructions for incorporating proper details for expected conditions and compatibility of their products with other envelope components.

Most manufacturers will, however, offer detailing suggestions and complete laboratory or site tests if required to ensure the inclusion of their product in a project. Manufacturers have technical resources available to them that are not immediately available to a designer or building owner.

Manufacturers should become involved in the design and testing process that can bring a project to a successful completion. Their intricate knowledge of their materials or systems and suggestions in termination and transition detailing should be consulted as a basis for preliminary design requirements.

Manufacturers often are capable of providing laboratory analysis and testing of their products under the proposed project conditions. This provides a means to determine acceptability of present design requirements or to suggest alternate designs. It also determines the compatibility of their products with other envelope components.

Manufacturers representing the various envelope components can become involved with final designing of a composite envelope to review and suggest revisions to ensure that the proposed detailing will work uniformly for all included products. When testing is completed at job sites or with laboratory mock-ups, manufacturers should be invited to review the tests and results and offer their opinions and suggestions.

Testing Deficiencies

Unfortunately, often envelopes still experience water infiltration after completion of testing. This can be caused by a variety of problems, including

- Insufficient termination and transition detailing in the test parameters
- Repair of defects found during testing by insufficient methods, including sealing with low-performance sealants
- Substitution of products, materials, or systems after the completion of testing
- Testing that does not reveal the long-term incompatibility of products, resulting in short life cycles or water infiltration
- Long-term weathering cycles not included in the testing
- Expected detrimental elements such as acid rain, road salts, ultraviolet weathering, and freeze–thaw cycling not included in the testing
- Actual field conditions not duplicated in the laboratory or mock-up testing, e.g., water at the actual job site containing chemicals detrimental to masonry admixtures and dry weather preventing proper curing of mortar
- Mock-ups constructed under laboratory conditions not possible at the site, including overall expertise of mechanics working on the actual building envelope
- Performance requirements not as demanding as required by actual job site conditions, e.g., wind loading (especially at upper building portions), thermal movement, and wear and durability required at locations such as loading docks
- Mock-ups not accounting for structural loading or settlement that will occur in actual building conditions

One major problem with envelope testing is a lack of standardized tests that allow review of typical transition and termination details. Further, there is a lack of testing designed specifically for waterproofing products. Tests supplied or used for waterproofing materials and systems are often those applied to roofing materials or other envelope systems.

The fact that water infiltration continues to plague building projects is evidence that either insufficient testing exists or tests and their results are often not considered seriously enough to warrant proper resolution of problems. All too often mock-ups are eventually made to pass requirements when sufficient quantities of sealants have been applied.

To ensure that test results are properly used, the following procedures should be followed:

- Any infiltration that occurs during testing should be carefully documented.
- Determination of the leakage cause should be completed before attempts at repair are completed.
- Repairs or redesigns incorporated into an envelope should be reviewed for compatibility with other components, and their life cycling must be adequate and equal to that of other components.
- Mechanics and supervisors should review proposed redesigns or repairs and be aware of their importance.
- After completion of repairs and redesign, envelopes should be retested to assure their adequacy.
- Manufacturers should be consulted to approve use of their materials and any redesign or repairs.
- Warranties should be reviewed to ensure they are not affected by repairs or redesigns.

Proper pretesting and resolution of design or construction flaws can prevent most of the problems that occur after completion of a building envelope. Successful testing must include adequate representative portions of all terminations and transitions incorporated into an envelope design. It is also mandatory that any leakage be reviewed and properly repaired to ensure the longevity and compatibility of the repair method.

Chapter

11

Information Resources

The following listing of manufacturers, government agencies, associations, and so on is a resource for acquiring information on many of the waterproofing systems, materials, and envelope components and processes provided in this book. The manufacturers are listed for information purposes only, and the list is not intended as recommendations of the companies or materials.

Most manufacturers will provide information on their products as well as generic or similar products. They are also excellent sources for resources for reviewing termination and transition detailing and compatibility of adjacent products in an envelope.

Below-Grade Waterproofing

Drainage protection courses

American Wick Drain Corporation
316 Warehouse Drive
Matthews, NC 28105
(704) 821-9300

Geotech Systems, Inc.
4100 Powers Court
Sterling, VA 22170
(703) 450-2366

Mirafi, Inc.
P.O. Box 240967
Charlotte, NC 28224
(704) 523-7477

Cementitious positive and negative systems

5 Star Products, Inc.
425 Stillson Road
Fairfield, CT 60430
(203) 336-7900

Anti-Hydro Company
265 Badger Avenue
Newark, NJ 07180
(201) 242-8000

Hey'Di American Corporation
2801 Crusader Circle
Virginia Beach, VA 23456
(804) 468-2200

Sto
P.O. Box 44609
Atlanta, GA 30336
(404) 346-0755

Thoro Systems Products
7800 NW 38th Street
Miami, FL 33166
(305) 592-2081

US Waterproofing, Inc.
425 Stillson Road
Fairfield, CT 06430
(203) 336-7970

Vandex
P.O. Box 1440
Columbia, MD 20144
(301) 964-1410

Fluid-applied systems

Anti-Hydro Company
265 Badger Avenue
Newark, NJ 07180
(201) 242-8000

Mameco International, Inc.
4475 East 175th Street
Cleveland, OH 44104
(216) 229-3000

Pecora Corporation
165 Wambold Road
Harleysville, PA 19438
(212) 723-6051

Tremco, Inc.
10701 Shaker Boulevard
Cleveland, OH 44104
(216) 229-3000

Hot-applied membranes

American HydroTech, Inc.
303 East Ohio Street
Suite 2120
Chicago, IL 60611
(313) 337-4998

Karnak Corporation
330 Central Avenue
Clark, NJ 07066
(201) 388-0300

W.R. Meadows
P.O. Box 543
Elgin, IL 60120
(312) 683-4500

Sheet systems

Carlisle Corporation
P.O. Box 7000
Carlisle, PA 17013
(717) 245-7000

Laurent Corporation
3047 Cannon Road
Twinsburg, OH 44087
(216) 425-7888

Nervastral, Inc.
2 Sound View Drive
Greenwich, CT 06830
(203) 622-6030

The Noble Company
614 Monroe Street
Grand Haven, MI 49417
(616) 842-7844

Polyguard
P.O. Box 755
Ennis, TX 75119
(214) 875-8421

W.R. Grace
Construction Products Division
62 Whittemore Avenue
Cambridge, MA 02140
(617) 876-1400

W. R. Meadows
P.O. Box 543
Elgin, IL 60120
(312) 683-4500

Clay systems

American Colloid Company
1500 West Shure Drive
Arlington Heights, IL 60004
(708) 392-4600

Applied Industrial Materials Corporation
421 East Hawley Street
Maundelein, IL 60060
(312) 949-6900

Paramount Technical Products, Inc.
9607 Girard Avenue South
Minneapolis, MN 55431
(612) 881-2368

Vapor barriers

Nervastral, Inc.
2 Sound View Drive
Greenwich, CT 06830
(203) 622-6030

St. Regis Paper Company
Laminated & Coated Products Division
55 Starkey Avenue
Attleboro, MA 02703
(617) 222-3500

Kendall
Construction & Engineering Products
One Federal Street
Boston, MA 02110
(617) 574-7000

Above-Grade Waterproofing

Clear repellents

Bostik Construction
Products Division
P.O. Box 8
Huntingdon Valley, PA 19006
(216) 674-5600

Chemical Corporation
P.O. Box 12599
North Kansas City, MO 64166
(816) 221-2712

Chemprobe Corporation
2805 Industrial Lane
Garland, TX 75041
(214) 271-5551

Euclid Chemical Company
19218 Redwood Road
Cleveland, OH 44110
(216) 321-7628

GE Silicones
260 Hudson River Road
Waterford, NY 12188
(518) 237-3300

Innovative Coatings
1100 N.W. 55th Street
Fort Lauderdale, FL 33309
(305) 938-9646

Okon, Inc.
6000 West 13th Avenue
Lakewood, CO 80214
(308) 232-3571

Sonneborn Building Products
7711 Computer Avenue
Minneapolis, MN 55435
(612) 835-3434

Tamms Industries
1222 Ardmore Avenue
Itasca, IL 60143
(312) 773-2350

Textured Coatings of America
5950 South Avalon Boulevard
Los Angeles, CA 90003
(213) 233-3111

Cementitious coatings

5 Star Products, Inc.
425 Stillson Road
Fairfield, CT 06430
(203) 336-7900

Bonsal
P.O. Box 24148
Charlotte, NC 28224
(704) 525-1621

Sto
P.O. Box 44609
Atlanta, GA 30336
(404) 346-0775

Tamms Industries
1222 Ardmore Avenue
Itasca, IL 60143
(312) 773-2350

Thoro Systems Products
7800 N.W. 38th Street
Miami, FL 33166
(305) 592-2081

Laticrete International, Inc.
1 Laticrete Park
Bethany, CT 06524
(203) 393-0010

Elastomeric coatings

Hydro-Shur
4000 Dupont Circle
Louisville, KY 40207
(502) 897-9861

Innovative Coatings
1100 N.W. 55th Street
Fort Lauderdale, FL 33309
(305) 938-9646

Textured Coatings of America
5950 South Avalon Boulevard
Los Angeles, CA 90003
(213) 233-3111

VIP Enterprises, Inc.
The Flood Company
P.O. Box 399
Hudson, OH 44236
(216) 650-4070

Deck coatings

3M
Adhesives, Coatings and Sealers Division
3M Center
St. Paul, MN 55144
(612) 733-1140

Crossfield Products
3000 East Harcourt Street
Compton, CA 90221
(213) 636-0561

Karnak Corporation
330 Central Avenue
Clark, NJ 07066
(201) 388-0300

Laticrete International, Inc.
1 Laticrete Park
Bethany, CT 06524
(203) 393-0010

Mameco International, Inc.
4475 East 175th Street
Cleveland, OH 44128
(216) 752-4400

MacNaughton Brooks
3637 Westen Road
Toronto, Canada M9L 1W1
(416) 741-3830

Pacific Polymers
12271 Monarch Street
Garden Grove, CA 92649
(714) 898-0025

Selby, Battersby & Co.
5220 Whitby Avenue
Philadelphia, PA 19143
(215) 474-4788

Tremco, Inc.
10701 Shaker Boulevard
Cleveland, OH 44104
(216) 229-3000

Clear deck sealers

3M
Adhesives, Coatings and Sealers Division
3M Center
St. Paul, MN 55144
(612) 733-1140

GE Silicones
260 Hudson River Road
Waterford, NY 12188
(518) 237-3330

Hodson Chemical Construction Corporation
955 North 400 West
North Salt Lake, UT 84054
(801) 292-3400

Hydrozo Coatings Company
1001 Y Street
Lincoln, NE 68501
(402) 474-6981

ProSoCo, Inc.
755 Minnesota Avenue
Kansas City, KS 66177
(913) 281-2700

Protected membranes

American Colloid Company
5100 Suffield Court
Skokie, IL 60077
(312) 966-5720

Laticrete International, Inc.
1 Laticrete Park
Bethany, CT 06524
(203) 393-0010

Mameco International, Inc.
4475 East 175th Street
Cleveland, OH 44128
(216) 752-4400

Polyguard
P.O. Box 755
Ennis, Texas 75119
(214) 875-8421

Tremco, Inc.
10701 Shaker Boulevard
Cleveland, OH 44104
(216) 229-3000

W.R. Grace
Construction Products Division
62 Whittemore Avenue
Cambridge, MA 02140
(617) 876-1400

W.R. Meadows
P.O. Box 543
Elgin, IL 60120
(312) 683-4500

Sealants

Acrylics

Innovative Coatings
1100 N.W. 55th Street
Fort Lauderdale, FL 30336
(404) 346-0755

VIP Enterprises, Inc.
The Flood Company
P.O. Box 399
Hudson, OH 44236
(216) 650-4070

Woodmount Products, Inc.
Huntingdon Valley, PA 19006
(215) 674-5600

Latex

Protective Treatments
3345 South 8th Road
Dayton, OH 45414
(513) 890-3150

Sonneborn Building Products
7711 Computer Avenue
Minneapolis, MN 55435
(612) 835-3434

Woodmount Products, Inc.
Huntingdon Valley, PA 19006
(215) 674-5600

Butyl

Protective Treatments
3345 South 8th Road
Dayton, OH 45414
(513) 890-3150

Woodmount Products, Inc.
Huntingdon Valley, PA 19006
(215) 674-5600

Urethane

Mameco International
4475 East 175th Street
Cleveland OH 44128
(216) 752-4400

Pecora Corporation
165 Wambold Road
Harleysville, PA 19438
(215) 723-6051

Quaker Sealants
8810 West 100th
Sapulpa, OK 74066
(918) 227-0176

Products Research & Chemical Corporation
410 Jersey Avenue
Gloucester City, NJ 08030
(609) 456-5700

Sonneborn Building Products
7711 Computer Avenue
Minneapolis, MN 55435
(612) 835-3434

Tremco, Inc.
10701 Shaker Boulevard
Cleveland, OH 44104
(216) 229-3000

Polysulfides

Products Research & Chemical Corporation
410 Jersey Avenue
Gloucester City, NJ 08030
(609) 456-5700

Sonneborn Building Products
7711 Computer Avenue
Minneapolis, MN 55435
(612) 835-3434

Thiokol Chemical Division
P.O. Box 1296
Trenton, NJ 08607
(609) 396-4001

Silicones

Dow Corning Corporation
Midland, MI 48686
(517) 496-4000

GE Silicones
260 Hudson River Road
Waterford, NY 12188
(518) 237-3330

Rhone-Poulenc, Inc.
CN 5266
Princeton, NJ 08543
(908) 297-9610

Sika Corporation
P.O. Box 297
Lyndhurst, NJ 07071
(201) 933-8800

Precompressed foam

Emseal Corporation
344 Mill Road
Stamford, CT 06903
(203) 322-3828

Will-Seal Construction Foams
3800 Washington Avenue North
Minneapolis, MN 55412
(612) 521-3555

Expansion Joints

Sealant systems

Nox-Crete Chemicals
P.O. Box 3764
Omaha, NE 68103
(402) 341-2080

Pecora Corporation
165 Wambold Road
Harleysville, PA 19438
(215) 723-6051

Tremco, Inc.
10701 Shaker Boulevard
Cleveland, OH 44104
(216) 292-5000

Sonneborn Building Products
7711 Computer Avenue
Minneapolis, MN 55435
(612) 835-3434

T systems

Nox-Crete Chemicals
P.O. Box 3764
Omaha, NE 68103
(402) 341-2080

Pecora Corporation
165 Wambold Road
Harleysville, PA 19348
(215) 723-6051

Harry S. Peterson Co.
4150 South Lapeer Road
Orion, MI 48359
(313) 373-8100

Tremco, Inc.
10701 Shaker Boulevard
Cleveland, OH 44104
(216) 292-5000

Expanding foam

Emseal Corporation
344 Mill Road
Stamford, CT 06903
(203) 322-3828

Will-Seal Construction Foams
3800 Washington Avenue North
Minneapolis, MN 55412
(612) 521-3555

Hydrophobic expansion seals

American Colloid
One North Arlington
1500 West Shore Drive
Arlington Heights, IL 60004
(312) 392-4600

de Neef America, Inc.
122 North Mill Street
St. Louis, MI 48880
(517) 681-5791

Sheet systems

Sika Corporation
P.O. Box 297
Lyndhurst, NJ 07071
(201) 933-8800

Bellows systems

Jeene Technology Corporation
1703 N.W. 38th Avenue
Lauderhill, FL 33311
(305) 486-3028

Seal Master Corporation
368 Martinel Drive
Kent, OH 44240
(216) 673-8410

Preformed rubber systems

Harry S. Peterson Co.
4150 South Lapeer Road
Orion, MI 48359
(313) 373-8100

Watson-Bowman Acme
95 Pineview Drive
Amherst, NY 14228
(716) 691-7566

Combination rubber and metal systems

Acme Highway Products
33 Chandler Street
Buffalo, NY 14068
(716) 876-0123

E-Poxy Industries, Inc.
14 West Shore Street
Ravena, NY 12143
(518) 756-6193

Vertical metal and stucco systems

Keene Corporation, Metal Products Division
Route 10, Box 14
Parkersburg, WV 26101
(304) 295-4581

Admixtures

Liquid and powder admixtures

5 Star Products, Inc.
425 Stillson Road
Fairfield, CT 06430
(203) 336-7900

Anti-Hydro Company
265 Badger Avenue
Newark, NJ 07180
(201) 242-8000

Bonsal
P.O. Box 24148
Charlotte, NC 28224
(704) 525-1621

Euclid Chemical Company
19128 Redwood Road
Cleveland, OH 44110
(216) 531-9222

Hey'Di American Corporation
2801 Crusader Circle
Virginia Beach, VA 23456
(804) 468-2200

Hodson Chemical Construction Corporation
955 North 400 West
North Salt Lake, UT 84054
(801) 292-3400

Lambert Corporation
20 North Coburn Avenue
Orlando, FL 32805
(305) 841-2940

Master Builders
23700 Chagrin Boulevard
Cleveland, OH 44122
(216) 831-5500

Tamms Industries
1222 Ardmore Avenue
Itasca, IL 60143
(312) 733-2350

Thoro Systems Products
7800 N.W. 38th Street
Miami, FL 33166
(305) 592-2081

Polymer concrete and overlays

Adhesives Technology Corporation
21850 88th Place South
Kent, WA 98031
(206) 872-2240

Bonsal
P.O. Box 24148
Charlotte, NC 28224
(704) 525-1621

General Polymers Corporation
9461 LeSaint Drive
Fairfield, Oh 45104
(513) 631-0649

Master Builders
23700 Chagrin Boulevard
Cleveland, OH 44122
(216) 831-5500

Nox-Crete Chemicals
P.O. Box 3764
Omaha, NE 68103
(402) 341-2080

Preco Industries, Ltd.
55 Skyline Drive
Plainview, NY 11803
(516) 935-9100

Remedial Waterproofing

Cleaning

HydroChemical Techniques, Inc.
P.O. Box 2078
Hartford, CT 06145
(203) 527-6350

ProSoCo, Inc.
755 Minnesota Avenue
Kansas City, KS 66177
(913) 281-2700

Cementitious coatings and tuck-pointing

5 Star Products, Inc.
425 Stillson Road
Fairfield, CT 06430
(203) 336-7900

Bonsal
P.O. Box 24148
Charlotte, NC 28224
(704) 525-1621

Hey'Di American Corporation
2801 Crusader Circle
Virginia Beach, VA 23456
(804) 468-2200

Master Builders
23700 Chagrin Boulevard
Cleveland, OH 44122
(216) 831-5500

Sto
P.O. Box 44609
Atlanta, GA 30336
(404) 346-0755

Tamms Industries
1222 Ardmore Avenue
Itasca, IL 60143
(312) 773-2350

Thoro Systems Products
7800 N.W. 38th Street
Miami, FL 33166
(305) 592-2081

Epoxy injection

Adhesives Technology Corporation
21850 88th Place South
Kent, WA 98301
(206) 872-2240

Multi-Chemical Products
2128 North Merced Avenue
South El Monte, CA 91733
(213) 686-0682

Sika Corporation
P.O. Box 297
Lyndhurst, NJ 07071
(201) 933-8800

Sto
P.O. Box 44609
Atlanta, GA 30336
(404) 346-0755

Chemical grouts

3M
Adhesives, Coatings and Sealers Division
3M Center
St. Paul, MN 55144
(612) 733-1140

Miscellaneous

American Society for Testing and Materials (ASTM)
1916 Race Street
Philadelphia, PA 19103
(215) 299-5400

Asphalt Roofing Manufacturers Association
6288 Mountrose Road
Rockville, MD 20852
(301) 231-9050

National Roofing Contractors Association
One O'Hare Center
6250 River Road
Rosemount, IL 60018
(312) 318-6722

Sealants & Waterproofers Institute
3101 Broadway, Suite 300
Kansas City, MO 64111
(816) 561-8230

Single Ply Roofing Institute
104 Wilmot Road
Deerfield, IL 60015
(312) 940-8800

Society of Plastics Industry, Inc.
Polyurethane Foam Contractors Division
1275 K Street N.W.
Washington DC 20005
(202) 371-5200

Glossary

Above-grade waterproofing The prevention of water intrusion into exposed structure elements through a combination of materials or systems. These materials are not subject to hydrostatic pressure but are exposed to weathering and pollutant attack.

Abrasive cleaning A cleaning method that incorporates an abrasive material such as sand to remove dirt, stains, and paint from existing substrates.

Accelerated weathering Controlled conditions applied in laboratory testing to condense greatly the weathering a waterproofing material would experience over a long life cycle. Test results are used to compare materials of different generic types or manufacturers.

Acrylic sealants Factory mixed one-component materials polymerized from acrylic acid. They are not used on joints subject to high movement due to their relatively low-movement capability.

Adhesion The ability of a waterproof material to bond to a substrate or other material during movement or stress.

Adhesive strength The ability of sealants to bond to a particular substrate, including adhesion during substrate movement.

Admixtures Materials added to masonry or concrete envelope components to enhance and improve in-place product performance.

Adsorption The surface absorption of water allowed by a waterproofing system. Testing for adsorption is carefully controlled under laboratory conditions to ensure uniform test results between different waterproofing materials.

Aliphatic Of or pertaining to materials such as urethane in which the molecular structure is arranged in open or straight chains of carbon atoms.

Alligatoring The cracking that occurs in a waterproofing material as a result of movement the material is not capable of withstanding. Alligatoring also occurs when substrate movement begins before final waterproofing material curing.

ASTM American Society for Testing and Materials, a nationally recognized and impartial society for the testing of building materials. Test results are used for comparisons among various types and manufacturers of materials.

Backing materials Backer rods and backing tape that prevent three-sided adhesion in joint design. When joints have insufficient depth for backer rod installation, tape is used at backs of joints, providing there is a firm substrate against which to install sealant. Backer rod is installed in joints where there is no backing substrate. Backing material also provides a surface against which to tool material and helps to maintain proper depth ratios.

Bag grouting Application of a cementitious waterproofing material to the entire face of a masonry envelope. The cementitious material is removed before it is completely set and cured by using burlap bags or stiff brushes. This is also referred to as face grouting because the entire face of the masonry facade is covered.

Base flashing Flashing that prevents water from wicking upward in capillary action in a masonry wall. *See* Flashing.

Bellows expansion systems Systems manufactured from vulcanized rubber into preformed joint sections. They are installed by pressurizing the joint cross section during adhesive curing, which promotes complete bonding to joint sides.

Below-grade waterproofing Use of materials that prevent water under hydrostatic pressure from entering a structure or its components. These systems are not exposed or subjected to weathering such as by ultraviolet rays.

Bentonite waterproofing systems Waterproofing materials composed primarily of montmorillonite clay, a natural material. Typically, bentonite waterproofing systems contain 85–90 percent of montmorillonite clay and a maximum of 15 percent natural sediments such as volcanic ash.

Blister A portion of a waterproofing material raised from the substrate because of negative vapor pressure or application over wet substrates.

Building envelope The combination of roofing, waterproofing, dampproofing, and flashing systems that act cohesively as a barrier, protecting interior areas from water and weather intrusion. These systems envelope a building from top to bottom, from below grade to the roof.

Butyl sealants Sealants produced by copolymerization of isobutylene and isoprene rubbers. Butyls are some of the oldest derivatives to be used for sealant materials.

Capillary admixtures Admixtures that react with the free lime and alkaline in a concrete or masonry substrate to form microscopic crystalline growth in the capillaries left by hydration. This crystalline growth fills the capillaries, resulting in a substrate impervious to further capillary action.

Caulking Joint sealing material appropriate for interior joints that exhibit little or no movement.

Chemical cleaning A cleaning method using a variety of chemical formulations to remove a number of substrate stains including paint, rust, and pollutants.

Chemical grouts Similar in application to epoxy injection repairs; however, these materials are manufactured from hydrophobic liquid polymer resins.

Chemical grouts are used for waterproofing cracks in a substrate and not for structural repairs.

Cohesive strength The ability of a material's molecular structure to stay together internally during movement. Cohesive strength has a direct bearing on elongation ability.

Cold joint Another name for construction joint, typically nonmoving in nature.

Construction joint A joint formed at the intersection of two separate concrete placements.

Control joint A joint in building or envelope materials that allows for substrate movement.

Counterflashing Flashing that is surface mounted or placed directly into walls with a portion exposed to flash various building elements, including roof flashings, waterproofing materials, building protrusions, and mechanical equipment, into the envelope. *See* Flashing.

Cure A process, whereby, through evaporation, heat, or chemical reactions, a waterproof material attains its final performance properties.

Curing agent A separate material applied immediately after application to waterproofing materials or substrates. Curing agents enhance curing time and properties.

Dampproofing A system that is resistant to water vapor or minor amounts of moisture and that acts as a backup system to primary waterproofing materials. Dampproofing materials are not subject to weathering or water pressure.

Delamination Separation of envelope materials from the applied substrate due to movement or improperly applied materials.

Detailing joints Joints required as a component or part of complete waterproofing systems. They are used for watertightness at building details such as pipe penetrations and changes in plane before application of primary waterproofing materials.

Differential movement A phenomenon that occurs because materials have individual characteristics of coefficients of movement that differ from surrounding envelope materials or systems. These differences will cause the materials to move at different rates during substrate movement.

Diffused quartz carbide This silica is suspended in a petroleum base of hydrocarbon solvent. Upon application, the sealer penetrates a substrate. The solvent evaporates and the silica bonds chemically with the substrate to form water-repellent properties.

DOT Department of Transportation federal and state agencies that monitor and regulate the transportation and disposal of hazardous waste.

Elasticity The measure of a sealant's ability to return to its original shape and size after being compressed or elongated. As with elongation, elasticity is measured as a percentage of its original length.

Elastomeric An adjective describing the ability of a waterproof material to return to its original shape and size after substrate movement during expansion or contraction.

Elongation The ability of a sealant to increase in length then return to its original size. Limits of elongation are expressed as a percentage of original size. A material with a 200 percent elongation, for example, is capable of stretching to double its original size without splitting or tearing. Also, the increase in length of an applied waterproofing material or system during expansion of the substrate.

EPA Environmental Protection Agency, a federal agency created to enforce and monitor regulations set by Congress relating to the environment, particularly those dealing with hazardous waste materials.

Epoxy injection The injection of low-viscosity epoxy materials into substrate cracks to restore the monolithic nature of the substrate. These can be used on wood, concrete, masonry, natural stone, or metal substrates. If additional substrate movement occurs, the epoxy may crack.

Expanding foam sealants These are composed of open cell polyurethane foam, fully impregnated with a manufacturer's proprietary product formulation. These products include neoprene rubbers, modified asphalts, or acrylic materials.

Expansion joint A break or joint in structural elements of a building that will continue to experience movement by thermal expansion and contraction.

Exposed flashings Flashings used in a variety of methods and locations. They can be an integral part of a system such as skylight construction or applied to provide protection between two dissimilar materials, including cap flashings, coping flashings, gravel stops, and edge flashings.

Face grouting *See* Bag grouting.

Flashing A material or system installed to redirect water entering through the building skin to the exterior. These are made from a variety of materials including noncorrosive metals and synthetic rubber sheet goods. Flashings are installed as backup systems for waterproofing or dampproofing systems. They are also used for waterproofing material transitions or terminations.

Floor flashing Flashing used in conjunction with shelf angles supporting brick or other facade materials. *See* Flashing.

Gunite Pneumatically applied small aggregate concrete or sand–cement mixtures, which are also referred to as shotcrete.

Head flashing Flashing installed above window head detail just below adjacent facing material that the window abuts. *See* Flashing.

Hydration The process of adding water to cement, sand, and aggregate, to form a paste that cures, hardens, and shrinks to create the finished concrete or masonry product. During curing, water leaves this paste through a process called dehydration, which causes formation of microscopic voids and cracks in concrete. Once formed, these voids allow water absorption through the material.

Hydraulic cement Frequently referred to as "hot patch" materials because of the heat generated during their extremely fast cure cycle. These materials are used to patch substrate cracks and small areas experiencing water leaking under hydrostatic pressure.

Hydrophobic expansion systems Systems combining hydrophobic resins with synthetic rubber to produce hydrophobic expansion seals. Hydrophobic refers to materials that swell in the presence of water. Thus, these materials require active water pressure to become effective water barriers.

Hydrostatic pressure Pressure applied to envelope materials by various heights of water at rest.

Isolation joints Joints that allow for any differential movement that will occur between two materials at junctures of these materials. For example, window frame perimeters require isolation joints when abutting other facade materials. These joints allow for differential movement at such locations as changes in structural components (e.g., spandrel beam meeting brick facing material).

Joint grouting Application of cementitious grout to all surfaces of existing mortar joints to repair the structure and waterproof effectively.

Laitance A thin layer of unbonded cement paste on concrete substrate surfaces that must be removed before waterproofing material application.

Latex sealants Typically, acrylic emulsions or polyvinyl acetate derivatives. Latex materials have limited usage for exterior applications. They are typically used for interior applications when a fast cure time is desired for painting.

Mechanic A person trained in the proper and safe application of a particular waterproofing system.

Modulus A measure of stress to strain; measured as tensile strength, expressed as a given percentage of elongation in pounds per square inch (lb/in^2). Modulus has a direct effect on elongation or movement capability.

Moisture content movement Movement caused by certain materials, particularly masonry substrates, swelling when subjected to wetting and subsequent drying. When this movement is calculated as an aggregate total of the entire facade area, it can be considerable.

Negative waterproofing systems Below-grade waterproofing systems applied to the interior or negative side of a structure, away from direct exposure to groundwater.

OSHA Occupational Safety and Health Administration, a federal agency that enforces workplace safety laws and regulations created by Congress.

Overlays Cementitious materials used for restoring deteriorated horizontal concrete substrates.

Parapet flashing Flashing installed at the base of a parapet, usually at ceiling level. It is also used on the roof side of parapets as part of roof or counterflashing. *See* Flashing.

Permeability The ability of a waterproofing material or substrate to allow the passage of water vapor through itself without blistering.

pH The chemical measurement of a substrate's alkalinity or acidity.

Polymer concrete A modified concrete mixture formulated by adding natural and synthetic chemical compounds known as polymers. Although the priority chemical compounds (polymers) vary, the purpose of these admixtures is the same. They provide a dense, high strength, low shrinkage, and chemically and water-resistant concrete substrate.

Polysulfides Materials produced from synthetic polymers of polysulfide rubbers. Polysulfides make excellent performing sealants for most joint use.

Polyurethane Any of various polymers that are produced by chemical reactions formed by mixing di-isocynate with a hydroxyl and are used in making flexible and rigid foams, elastomers, and resins. Many polyurethanes are moisture-cured materials reacting to moisture in atmospheric conditions to promote curing. Other polyurethanes are chemically curing mixtures.

Positive waterproofing systems Waterproofing systems applied to substrates side with direct exposure to water or a hydrostatic head of water.

Pot life The length of time a waterproof material or system is workable or applicable after having been activated.

Poultice An absorbent material such as talc or fuller's earth that is applied to the envelope substrate to remove dirt and staining. The poultice absorbs the staining and dirt into itself, and then is removed by water pressure cleaning.

Primer A separate material, usually in liquid form, applied to a substrate before actual waterproofing material application. Primers enhance adhesion properties of the waterproofing system.

Protected membrane A membrane applied between a structural slab and topping slab or other top layer protection such as tile. The topping slab or protection slab protects the membrane from weathering and traffic wear. This is also referred to as a sandwich membrane.

Reglet A formed or sawn groove in substrate (usually concrete) providing a transition point for two adjoining waterproofing systems.

Remedial flashings Flashings typically surface mounted and applied directly to exposed substrate faces. These can include a surface-mounted reglet for attachment. They do not provide for redirecting entering water. Only by dismantling a wall or portion thereof can remedial through-wall flashings be installed.

Roofing That portion of a building that prevents water intrusion in horizontal or slightly inclined elevations. Although typically applied to the surface and exposed to the elements, roofing may also be internal, or "sandwiched" between other building components.

Sandwich membrane *See* Protective membrane.

Sealant A material applicable to exterior building envelope joints. Sealants are capable of withstanding continuous joint movement during weathering conditions without failing.

Shear movement Lateral movement in a substrate.

Sheet expansion systems Systems manufactured from neoprene or hypalon rubber sheets. Joint expansion and contraction is made watertight by installing these materials in a bellows or loop fashion.

Shelf angle Steel angle extrusion used over envelope openings to support masonry and precast and other cladding materials.

Shelf life The maximum time packaged and unopened waterproofing materials can remain usable.

Shore hardness A measure of resistance to impact using a durometer gage. This property becomes important in choosing sealants subject to punctures or traffic, such as horizontal paver joints. A shore hardness of 25 is similar to a soft eraser; a hardness of 90 is equivalent to a rubber mallet.

Shotcrete *See* Gunite.

Silanes Water repellents that contain the smallest molecular structures of all silicone-based materials. The small molecular structure of the silanes allows the deepest penetration into substrates. Silanes must have silica present in substrates for the chemical action providing water repellency to take place. These materials therefore are inappropriate for substrates such as wood, metal, and limestone.

Silicone water repellents Water repellents manufactured by mixing silicone solids (resins) into a solvent carrier. Most manufacturers base their formulations on a 5 percent solids mixture in conformance with the requirements of federal specification SS-W-11OC.

Silicones Silicone sealants are derivatives of silicone polymers produced by combining silicon, oxygen, and organic materials. Silicones have extremely high thermal stability and are used as abrasives, lubricants, paints, coatings, and synthetic rubbers. *See* Sealant.

Sill flashing Flashing installed beneath window or curtain wall sills. *See* Flashing.

Siloxanes Silicone masonry water repellents produced from the CL-silane material. Siloxanes are manufactured in two types—oligomerous (meaning short chain of molecular structure) alkylalkoxysiloxanes and polymeric (long chain of molecular structure) alkylalkoxysiloxanes. Most siloxanes produced now are oligomerous due to a tendency for polymeric products to remain wet or tacky on the surface, attracting dirt and pollutants.

Sodium silicates Materials that react with the free salts in concrete, such as calcium or free lime, making the concrete surface more dense. Usually these materials are sold as floor hardeners.

Structural movement Substrate movement caused by the curing process in concrete during settlement and/or structural loading of a building.

Substrate Structure or envelope components to which waterproofing materials or systems are applied.

T-joint system A sealant system reinforced with metal or plastic plates and polymer concrete nosing on each side of the sealant. This system derives its name from a cross section of the joint, which is in the shape of a T.

Tackiness Stickiness of a waterproofing material's exposed surface after installation or during its final curing stage.

Tensile strength The ability of a waterproofing material to resist being pulled or stretched apart to a point of failure.

Thermal movement Movement, either expansion or contraction, caused by temperature changes.

Thiokol Trademark of the first commercial synthetic elastomeric produced by Thiokol Chemical Company.

Tooling The means of finishing mortar or sealants that have been applied into envelope joints.

Ultraviolet A form of light energy positioned in the spectrum of sunlight beyond violet, the limit of visible light, that is an especially detrimental weathering condition to above-grade waterproofing materials and envelope components.

Viscosity The chemical property of liquid waterproofing materials to resist a change in shape.

Water cleaning A building cleaning method incorporating water by pressure, soaking, or steam to remove dirt and pollutants from a substrate.

Waterstops Preformed materials placed between construction joints in separate concrete placements to prevent passage of water between the joints that forms at this intersection.

Weep hole Opening in a masonry wall or sealant that allows the escape of water entering envelopes to the exterior. Exiting water is collected by a combination of dampproofing and flashing materials.

Index

ABOUT THE AUTHOR

Michael T. Kubal is vice-president of the Chas. H. Tompkins Co., Washington, D.C., a subsidiary of J.A. Jones Construction Company. He has extensive experience as a project manager on multimillion dollar building projects, as well as a supervisor of renovation and historical restorations. With more than 15 years of direct experience in waterproofing/restoration, Mr. Kubal has acted as an expert witness, consultant, and lecturer on waterproofing and related problems.